4/04

USING ANIMALS FOR ENTERTAINMENT

Titles in the series:

EXPLORING ANIMAL RIGHTS AND ANIMAL WELFARE

VOLUME 3

USING ANIMALS
FOR ENTERTAINMENT

GREENWOOD PRESS

Westport, Connecticut • London

Library of Congress Cataloging-in-Publication Data

Creative Media Applications
Trumbauer, Lisa, 1963–
    Exploring animal rights and animal welfare / Lisa Trumbauer.
      v. cm.
    Contents: v. 1. Using animals for food — v. 2. Using animals for research — v. 3. Using animals for entertainment — v. 4. Using animals for clothing.
    Includes bibliographical references and index.
    ISBN 0-313-32245-7 (set) — ISBN 0-313-32246-5 (v. 1) — ISBN 0-313-32247-3 (v. 2) — ISBN 0-313-32248-1 (v. 3) — ISBN 0-313-32249-X (v. 4)
      1. Animal welfare — Juvenile literature. 2. Animal rights — Juvenile literature. 3. Human-animal relationships — Juvenile literature. 4. Humane education — Juvenile literature. [1. Animals — Treatment. 2. Animal rights. 3. Human-animal relationships.] I. Title. II. Series.

HV4712.T78 2002
179.3 — dc21   2002075303

British Library Cataloguing in Publication Data is available.

Library of Congress Catalog Card Number: 2002075303
ISBN: 0-313-32245-7 (set)
       0-313-32246-5 (Vol. 1)
       0-313-32247-3 (Vol. 2)
       0-313-32248-1 (Vol. 3)
       0-313-32249-X (Vol. 4)

First published in 2002

Greenwood Press, 88 Post Road West, Westport, CT 06881
An imprint of Greenwood Publishing Group, Inc.
www.greenwood.com

Printed in the United States of America

∞™

The paper used in this book complies with the Permanent Paper Standard issued by the National Information Standards Organization (Z39.48–1984)

10 9 8 7 6 5 4 3 2 1

Photo Credits:
Cover Image: AP/Wide World Photographs
Alan Barnett: Page 6
AP/Wide World Photographs: Pages 4, 6, 11, 14, 19, 20, 24, 32, 38, 46, 48, 50, 56, 58, 62, 67, 72, 76, 79, 84, 87, 92, 97, 98, 102, 106

A Creative Media Applications, Inc. Production

Writer: Lisa Trumbauer
Design and Production: Alan Barnett, Inc.
Editor: Matt Levine
Copyeditor: Laurie Lieb
Proofreader: Betty C. Pessagno
AP Photo Researcher: Yvette Reyes
Consultants: Sheryl Dickstein-Pipe, Ph.D. and
    Stephen Zawistowski, Ph.D., Certified Applied Animal Behaviorist

EXPLORING ANIMAL RIGHTS AND ANIMAL WELFARE

VOLUME 3

USING ANIMALS FOR ENTERTAINMENT

CONTENTS

Introduction

When you hear the word *entertainment*, you may think of actors and actresses in the movies or on television. Actors and actresses pretend to be characters in stories that we watch on television, on movie screens, and in theaters. Other human entertainers include singers, comedians, mimes, magicians, circus performers, and talk-show hosts.

People enjoy being entertained for a variety of reasons, one of which is to witness something different from what they see in their ordinary lives. What can be more different than being entertained by animals? Sometimes people are entertained by simply watching animals in the wild or in television documentaries. Animals and their natural behaviors can be fascinating to observe. Other times, animals are trained or used in order to entertain people. For example:

- Animals are featured in movies and television shows.

- Animals are trained to perform in circuses.

- Animals are trained to compete against each other, as in horse or dog racing.

- Animals are used to compete against people, as in rodeos or bullfights.

- Animals are viewed and judged, as in dog shows or 4-H fairs.

1

Even animals in zoos are a form of entertainment. Watching a lion basking in the sun or seals leaping from the water provides sights that most people do not experience in their everyday lives. Seeing a wild animal in the confines of a zoo is thrilling for many zoo visitors. It opens a door to a world that they could otherwise glimpse only if they traveled to the country where the animals actually lived in the wild.

WHAT'S THE PROBLEM?

Using animals for the sake of entertainment sounds like fun for the people who watch them and the animals alike. However, not all people believe that using animals for entertainment is necessary. In fact, some people believe that using animals for entertainment is *inhumane,* or cruel. The treatment of performing animals has become an important issue for many *animal rights* and *animal welfare activists,* or people who try to better the welfare of animals.

Animal groups question many aspects of a performing animal's life, including how the animal is trained, how it is treated before and after performances, and the condition of the animal's home. For example, once an elephant or lion leaves a circus ring, where does it sleep and how is it treated? Is it fair to keep an animal in a cage at a zoo, when the animal would normally roam freely in the wild? Is it against the nature of dolphins and seals to learn tricks that they must perform on cue before an audience? And if it is against the animal's nature, is it therefore inhumane?

The philosophy of some people who champion animal rights is that the animals should not be made to do anything that they would not do naturally. However, many animals actually do enjoy performing. Like human entertainers, some animals take to performing naturally, and their personalities shine when the spotlight is on them. Dog owners who enter their pets in agility contests describe their dogs as excited and happy before events. People who train dolphins in an aquarium describe the dolphins as eager and willing to please their trainers.

Problems creep in when animals are pushed beyond their limits—for example, when they are expected to perform even though they are too tired. Other problems arise when saving money outweighs animal welfare. For example, building smaller cages is less costly than building bigger cages, but bigger cages may be better for the animals.

Another issue for animal rights and animal welfare groups is animal emotions. Although scientists disagree about the extent of an animal's ability to experience emotions, many agree that animals do have feelings on some level. How then do animals feel when they are forced to become entertainers or when they spend their lives in small cages? Should humans who deal with animals consider these emotions, or should humans merely provide the animals with enough food and water?

A JUMBO EXAMPLE

The true story of Jumbo the elephant illustrates the problems that arise when an animal is taken out of its natural habitat for the purpose of entertainment. Jumbo was captured as a baby in Africa in the mid-1800s. He was bought by an animal dealer—someone who bought and sold animals to make a profit—who then sold him to a zoo in Paris, France. Jumbo was not properly cared for at the Paris zoo, and he became too thin. Eventually, the Paris zoo traded Jumbo to the Royal Zoological Gardens in London, England.

If you believe that animals experience emotions, you can only imagine the confusion and fear that the young elephant must have felt. First, he had been torn away from his homeland and family and transported to an unfamiliar environment. Now, his life was once again uprooted as he changed zoos from Paris to London.

In the mid-1950s, college students from Tufts University in Medford, Massachusetts, place coins in the trunk of Jumbo the elephant for good luck. Jumbo was stuffed and displayed on campus at Barnum Hall until 1975, when the hall burned down.

However, the move to London was for the best. A man named Matthew Scott began to care for Jumbo, and the elephant grew and thrived. He showed his appreciation to Scott by wrapping his trunk around the man's waist in an affectionate hug. Eventually, Jumbo grew to become the largest animal in captivity in the world. He stood 11 feet (3.3 meters) tall, and he reportedly weighed 7 tons (6.3 metric tons). He lived at the London zoo for seventeen years, becoming one of its most popular attractions.

Jumbo was not always a pleasant elephant. Sometimes at night, he appeared to have tantrums. He banged his massive body against the walls of his cage and thrust his long, ivory tusks against the bars. In the process of his tantrums, Jumbo broke his tusks on the bars. Perhaps he was trying to tell his keepers something.

The London zoo was nervous about Jumbo's behavior. What would happen if the 7-ton elephant broke out of his cage and ran through the streets of the city? He might hurt not only himself, but others, too. The zoo felt that keeping Jumbo was too much of a risk. When he was twenty-one years old, Jumbo was sold to a man in the United States named Phineas T. Barnum, who wanted Jumbo for his circus.

Once again, Jumbo was on the move. This time, though, he was allowed to take with him something from his home— Matthew Scott, his trainer. Jumbo was shipped across the Atlantic Ocean to America, and for three and a half years, he performed in Barnum's circus. He became so well known in America that his name, Jumbo, eventually came to mean the same as the word *big*.

Then, in 1885, Jumbo's life came to a tragic end. In order to move the circus from town to town, both animals and people traveled by train. The animals were being loaded onto a train when another train came rushing down the track. Jumbo was hit by the train and died.

Jumbo's life probably would have been very different if he had been allowed to remain in Africa. Animal rights activists would probably say that Jumbo would have been happier if he had lived a more normal life with his elephant family in his natural habitat. However, others might argue that Jumbo's life was better spent as a zoo and circus attraction. Because people were able to actually see Jumbo, they perhaps gained a greater appreciation for elephants the world over.

⟳ CLOSER TO HOME

Jumbo and Matthew Scott were an example of how an animal and a human can grow to appreciate and care for one another. Although the relationship between a person and an elephant is a bit exotic, examples of the bond between humans and animals are apparent every day in the relationships between

Cats such as this one have been kept as pets for more than 5,000 years. Modern pet cats are thought to be descendants from African cats first domesticated in ancient Egypt.

people and their pets. Dogs and cats are, of course, the most common animals that are pets, but people also own birds, rabbits, hamsters, turtles—even snakes and lizards. A visit to a pet shop reveals the variety of animals that people can choose for household pets.

The question then arises—if it's okay to keep some animals, like cats and dogs, in the care of people, then why is it not okay to keep all animals in this way? After all, animals—be they cats or dogs, lions or elephants—are all animals, right?

Some animal rights activists believe that all animals should live as nature intended. "In a perfect world, animals would be free to live their lives to the fullest: raising their young, enjoying their native environments, and following their natural instincts," says the animal rights group People for the Ethical Treatment of Animals (PETA).

However, many animals, such as dogs and cats, are *domesticated* animals. That means that through generations, dogs and cats have become used to living with people, and their babies are instinctively used to people also. Animals in the wild are *wild* animals. They are not used to living with people. They would much prefer to live in the wild than in human-made confinement. Although a wild animal may become *tame,* or used to living with people, its babies will, in most circumstances, be born wild; they will not be tame.

Both domesticated and wild animals are used for entertainment. Some people feel that domesticated animals are better suited to entertainment because they are more used to being around people. The treatment of the animals, whether wild or domesticated, should be a concern for the people who work with these animals that entertain. This book sets out to explore how animals are used for entertainment, how animal performers are treated, and how animal activists view such animals in entertainment.

CHAPTER 1

Animal Stars

Lights, camera, action!

These three words are traditionally called out when filming begins on a movie. Movies and television shows are enormously popular forms of entertainment. Ever since they began, animals have been gracing movie and television screens.

THE EARLY DAYS OF MOVIES

The movie industry began about a hundred years ago. Early movies did not look anything like the movies of today. They were in black and white, they had no sound, and the action on the screen moved very fast.

Early moviemakers realized that people enjoyed watching animals on film. Some of the animals shown in early movies were not real animals. Instead, actors wore animal costumes, or models were made of the animals. These make-believe animals usually did not fool the audiences. They wanted to see the real things.

Moviemakers started keeping animals at the film studios—both common animals, like cats, dogs, and horses, and more exotic animals, like lions and elephants. Keeping animals at the movie studios became very expensive. Special companies were then formed to provide movie studios with the animals they needed.

Model Animals

It isn't always possible to use real animals in movies, so moviemakers sometimes use models, or *miniatures*. For example, the movie *King Kong* (1933) needed a gorilla of giant proportions. Miniatures and special filming techniques brought King Kong to life. Today, animals can be re-created using computer technology. Computer animation created the dinosaurs in the *Jurassic Park* movies.

DOG DAYS

One of the most popular animals in movies and television is the dog. The characters on the television show *Frasier* are often upstaged by a Jack Russell terrier named Eddie. Dogs are also used in commercials, and they have been central figures in many motion pictures.

One of the first dogs to act in the movies was Etzel, more commonly known by his screen name, Strongheart. He was a German shepherd who had many heroic adventures in movies during the 1920s. In 1925, another canine star was on the rise—Rin Tin Tin, who was also a German shepherd. Rin Tin Tin and his movies became so popular that the money they made saved the movie studio that created them from going bankrupt.

Perhaps the most famous dog star of all was Lassie, a collie. Lassie made her debut in the movie *Lassie Come Home* (1943). Lassie's trainer was a man named Rudd Weatherwax. People enjoyed watching the first Lassie movie so much that more Lassie movies were made, as well as Lassie television shows.

In the 1970s, a big box-office hit was the movie *Benji* (1974), starring a smallish, scruffy-looking dog. Benji was different from Rin Tin Tin and Lassie because he wasn't *purebred*—that is, he wasn't a specific breed of dog. Instead, he was a mixed breed, found at an animal shelter. Benji, whose real name was Higgins, was on the television show *Petticoat Junction* for seven years before he moved into the movie spotlight.

OTHER DOMESTICATED ANIMALS

Dogs are domesticated animals. Other domesticated animals include cats, horses, and pigs, all of which have enjoyed movie or television fame.

A cat named Orangey found his way to stardom with the aid of an animal casting company called Frank Inn. Orangey starred in several movies throughout the 1950s and 1960s. A Disney movie called *That Darn Cat* (1966) starred a cat who helped the

Young actor Roddy McDowall sits beside a collie named Lassie in the 1943 movie Lassie Come Home.

11

Battle of the Sexes

The name *Lassie* implies that Lassie was a female dog. But most Lassies were actually played by male dogs. Here's the reason why: Female collies shed their fur more than male collies. The first Lassie was supposed to be a female dog, but because she was shedding, she was replaced by a male dog—Rudd Weatherwax's dog, Pal. All Lassies were descendants of Pal.

FBI. In the 1990s, the television show *Sabrina, the Teenage Witch* featured a talking black cat. Some scenes were filmed with a live cat, and others were filmed with a mechanical one.

Horses are another movie favorite. Dozens of horses may be used in a western. A well-known cowboy named Roy Rogers had a horse named Trigger. Trigger lived for thirty-three years and starred in eighty-seven movies and over a hundred television episodes. *Mr. Ed* was a television show about a talking horse. Other popular movies that featured horses include *My Friend Flicka* (1943) and *National Velvet* (1944).

In 1995, the movie *Babe*, about a talking pig trying to find his place on a farm, was very successful. It was nominated for the Academy Award for Best Picture, and it won the Golden Globe Award for Best Motion Picture, Musical or Comedy. It was followed by a sequel, *Babe: Pig in the City* (1998).

WILD ANIMALS

Unlike domesticated animals, wild animals need a bit more care on the movie set. That hasn't stopped filmmakers from using wild animals. Lions were the focus of the movie *Born Free* (1966), featuring Elsa the lioness. Elephants have appeared in numerous movies, including *The Jungle Book* (1942), *Jumbo* (1962), and *Larger Than Life* (1996). Cheetah was Tarzan's chimpanzee friend in the Tarzan movies, and chimps were an integral part of the movie *Project X* (1987). Bart, a well-trained grizzly bear, appeared in such movies as *The Clan of the Cave Bear* (1986), *The Bear* (1989), and *The Edge* (1997).

Sea creatures have also appeared in movies. *Flipper* (1963), a popular movie and television show in the 1960s, had a dolphin in the starring role. In 1993, the movie *Free Willy* explored the relationship between a young boy and a killer whale. In 1997, the movie industry combined people's fascination with dolphins and their love of dogs by pairing the two in *Zeus and Roxanne.*

Other animals have appeared in films, as well, including rats, bees, snakes, piranhas, and crocodiles. These movies often portray the animals as fierce killers, bent on doing harm to people. However, due to the unpredictable behavior of animals on a movie set, the only creatures being harmed may be the animals themselves.

Money Makers

Just like human actors, animal actors earn money. (The money, of course, goes to the person who owns the animal.) According to an agency that specializes in casting animals for movies and television shows, these salaries are earned by animal actors:

Animal	Cost per Day
Horse	$150
Dog	$300
Pig	$300
Chimp, adult	$800
Large alligator	$1,000
Grizzly bear	$2,500
Elephant	$2,500

⊚ SAFE ON THE SET?

A big concern for many people, both on and off a movie or television set, is the safety of animals involved in moviemaking. Until the 1940s, not much thought was given to animal safety. Sometimes, animals even died during filmmaking. One movie script called for a lion to be killed. An old lion was brought in for that purpose, and the lead actor stabbed the lion to death. In another movie, a horse and an actor jumped off a cliff. The actor survived, but the horse did not.

Babe the pig is shown in a scene from the 1998 movie Babe: Pig in the City.

After the latter incident, the American Humane Association (AHA) stepped in. In 1940, it opened offices in Los Angeles, California, the heart of the filmmaking industry, for the specific purpose of monitoring the treatment of animals on sets for movies and, eventually, television. However, critics of the AHA complain that the group does not monitor what happens to the animals before or after a performance, only during the time that they are on the set. These critics say that while the animals might not be mistreated during their performances, they may be mistreated during training or after performing.

One way to prevent animals from being harmed is to use mechanical ones. In early motion pictures, horses that were required to fall in battle scenes were sometimes actually tripped by wires, hurting and possibly injuring them. In the 1990 film *Dances with Wolves,* many buffalo fall during a buffalo hunt. These buffalo, however, were mechanical models, not live buffalo. In this way, no animals were harmed during filming.

Today, animals in the movie and television spotlights are certainly treated better than they once were. With improvements in moviemaking special effects, the use of real animals in this area of entertainment could one day be obsolete.

Before They Can Be Stars

Before a film that involves animals even begins shooting, the American Humane Association (AHA) is involved. First, the AHA reviews scripts, pointing out any areas that could be dangerous to the animals. Next, the AHA visits the movie or television set to see where and how the animals will be cared for, as well as props and other movie devices that might pose threats to the animals. Finally, the AHA reviews the movie and its treatment of animals while filming. "We ensure that stunts, safety measures, camera angles, special effects, and even lighting, make-up and costumes for animal actors receive the same planning and consideration as for human stars," states the AHA on its Web site.

CHAPTER 2

At the Zoo

Imagine that you're an animal in the wild, minding your own business. Perhaps you're resting in the shade of a tree, or maybe you're grazing on a grassy plain. Your only concern is your search for food and perhaps the welfare of your young.

Suddenly, the quiet of the wilderness is broken by a strange animal that walks on two legs. You try to get out of the animal's way, only to find that this strange creature is chasing you. Before you know it, you've been captured and placed inside a box with metal bars.

For the rest of your life, this is your home. No longer will you feel the grass beneath your feet. No longer will you roam freely and lounge in the sun or shade. A box no bigger than yourself is the only life you now know. That, and the other strange, two-legged animals that continually stare, point, and laugh at you.

For centuries, this was the plight of most animals captured and put on display at zoos. Not many people understood the needs of wild animals. In the 1800s, people began to take more of an interest in *zoology*—the study of the animal kingdom. In the late 1800s and early 1900s, many humane societies were

Word Origins

The word *zoo* comes from the Greek word *zoion,* which means "living being." The suffix *-ology* means "the study of." So *zoology* means "the study of living beings."

formed to monitor the treatment of animals. Today, many zoos and their facilities have greatly improved. Here is a brief history of zoos and the humane efforts involved in zoos.

THOUSANDS OF YEARS AGO

People have put animals on display for thousands of years. Records from 1500 B.C.E. tell of an Egyptian pharaoh who had a collection of animals. Alexander the Great, who ruled over most of the known world during the fourth century B.C.E., owned a variety of animals, including monkeys, elephants, and bears. In ancient Rome, from 27 B.C.E. to A.D. 476, Roman emperors felt that the keeping of large and exotic animals, such as lions, tigers, bears, and elephants, demonstrated their wealth and power. These animals were not only on display, but were used as entertainment in the *coliseums*—large, stadiumlike arenas. The animals fought gladiators, fought each other, or merely pounced on unarmed people.

Collecting animals as a symbol of wealth and power continued until about the 1700s. Then the power of kings, queens, and emperors began to wane. Soon, animal collections —*menageries*—were enjoyed not only by the wealthy, but by the common people as well.

THE BIRTH OF ZOOLOGICAL GARDENS

In the 1800s, the word *zoology* was combined with the word *garden,* creating the term *zoological garden.* The term referred to a place in which people could observe animals. Strolling through a zoological garden, people could view rows of cages holding bears, elephants, monkeys, lions, and other animals that do not commonly live in big cities.

What people failed to realize back then, however, was that they were not viewing the true lives of the animals. Most animals in zoological gardens had been taken out of their natural habitats and placed in a very unnatural setting of hard tile or dirt floors and metal bars. Animals that normally would not *cohabit*, or live together, were often displayed in cages side by side. Animals that usually lived within social groups, like monkeys, were often exhibited in cages by themselves. And, of course, the natural plants and animals that were a part of the animals' diets were missing. However, new ideas about keeping and exhibiting animals were on the horizon.

THE NEW ZOOS

A man named Dr. William Camac in Philadelphia, Pennsylvania, felt that his city could benefit from a zoo. It took many years of planning and fund-raising, but in 1874, the Philadelphia Zoo finally opened. Here, people were greeted with more than just a

Polar bears at the San Diego Zoo in California, shown here in August 1996, are housed in an open enclosure designed to be like their natural environment. Visitors ride past on a double-decker tour bus.

Lucky, right, a white lioness cub, is shown with her two brothers in May 1997.
These lions were born in captivity in the Columbus Zoo in Columbus, Ohio.

disorganized array of exotic wildlife. Buildings such as the monkey house and the bird house had been constructed specifically for each type of animal. Visitors were able to observe prairie dogs in the prairie dog village and watch sea lions frolicking in their own pools. Dr. Camac also hired full-time help to care for the animals.

In another revolutionary move, the Philadelphia Zoo opened a research facility in 1900. The purpose of the facility was to enable scientists to learn about the animals that the zoo housed. Because of the research facility, the Philadelphia Zoo was a step in the evolution of zoos from primarily a form of entertainment to an establishment for education, preservation, and conservation. It was an important change and one that other zoos would soon follow.

The Wildlife Conservation Society

The Wildlife Conservation Society (WCS) began in 1895. Back then, it was called the New York Zoological Society. The society was established to monitor the care and conservation of animals in New York City zoos. Today, it works with fifty-three countries around the world to protect natural habitats. As stated on its Web site: "We uniquely combine the resources of wildlife parks in New York with field projects around the globe to inspire care for nature, provide leadership in environmental education, and help sustain our planet's biological diversity." What this means is that the society tries to get people to care about nature. It also tries to help maintain the variety of plants and animals (biological diversity) that live on Earth. It does these things through its zoos and through projects that it supports around the world. The WCS is one example of how zoos are working to right the wrongs that people have done to animals in the past.

Zoo Atlanta

One example of the ways in which zoos have improved over the years is Zoo Atlanta in Georgia. For decades in the past, the gorillas at the zoo were housed in cages, rarely getting the chance to go outside. Then in 1984, a man named Terry Maple became the director of the zoo. He was a *primatologist*, a scientist who specializes in the study of primates. *Primates* are a group of animals that includes apes, monkeys, and humans. The living conditions of the gorillas deeply troubled him, and he spent several years working on a new living space for the animals. In 1988, the Ford African Rain Forest opened. Now the gorillas have a chance to live in a habitat that closely resembles their own.

Although conditions for the animals had improved, they were still deprived of many things—mostly a sense of their own habitats. The life that most zoo animals faced was not pleasant. The animals merely existed in cramped cages, day after day after day. Although people wanted to see these exotic, amazing creatures, no one was too concerned with what the animals really needed, such as the companionship of members of their own species or stimulating activities. After all, they were merely animals. Other than food, water, and shelter, what other needs could they possibly have?

A TIME OF CHANGE

One of the things that initiated a change in the way zoo animals were treated was a lack of animals to be captured. Because so many animals had been either killed or taken away from their natural habitats, animals in the wild were becoming harder to find. Zookeepers soon realized that they had to start taking better care of the animals in their zoos. This would ensure that the animals would live longer and reproduce. One way to obtain this goal was to try to re-create the animals' natural habitats.

A man in Germany had a unique vision. Carl Hagenbeck wanted to show animals not as captured creatures behind bars, but as free beings roaming in natural settings. Construction on Hagenbeck's zoo began in 1890. In order to contain the animals in specific areas, deep moats were dug around the areas and then covered with plants. The animals could not cross these moats, and they soon learned to live within their new, cageless confines. Called *naturalistic exhibits*, the attractions in Hagenbeck's zoo included lionesses relaxing with their cubs and herds of zebras and antelope grazing.

Although some people debated the merits of these new zoos, others thought that they were a great idea. Hagenbeck's sons were hired to help design the zoos in Detroit, Michigan, and Cincinnati, Ohio, and many other zoos were based on the idea of naturalistic exhibits. Not only were these new zoo designs better for the animals, but they helped people learn more about animals and their behaviors in the wild.

GREATER IMPROVEMENTS

It had taken centuries for the plight of animals in zoos to improve. In the 1900s, people became more aware of animals and animal welfare, through new discoveries in zoological science and through greater empathy with all living creatures.

Perhaps *empathy*—the ability to understand the thoughts and feelings of others— is what is needed most when considering the animals in a zoo. Scientists who developed a keener understanding of animals, their behavior, and their habitats came to realize that although food may keep an animal alive, food alone does not create a happy or productive animal. Zookeepers also learned that people who visit a zoo would much rather see an active animal that is content in a natural setting than a caged animal that seems depressed and gets little care.

Bula, a lion cub, snuggles with his mother Claserie in their habitat at Zoo Atlanta in Atlanta, Georgia, in May 1996.

Roadside Attractions

Although many large zoos around the country have made great improvements in their facilities and their treatment of animals, many smaller zoos have not. Many small zoos are also known as roadside zoos, because they are built along highways, hoping to entice passersby to visit. Many of these roadside zoos still exhibit animals as they were shown in the old animal menageries, living in barren cages instead of lush landscapes that resemble their habitats.

Many zoos have worked hard to re-create animals' natural habitats. For example, Zoo Atlanta in Atlanta, Georgia, has the remarkable Ford African Rain Forest, in which gorilla families live in a habitat much like the one gorillas are accustomed to in the wild. At JungleWorld at the Bronx Zoo in New York, monkeys, otters, and gharials (reptiles similar to crocodiles) live in as natural a setting as possible. In North Carolina, incredible birds inhabit the R.J. Reynolds Forest Aviary. In California, impalas leap across a grassy field at the San Diego Wild Animal Park, which, along with the world-famous San Diego Zoo, is part of the Zoological Society of San Diego.

Zoos have come a long way since they were merely "animal collections," or menageries. In fact, many zoos today are considered wildlife conservation centers. The zoos strive not only to share their animals with people, but to learn more about the animals in order to preserve each species. Even so, the question begs to be asked: How do these "realistic" displays appear from the animals' own perspective? Would the animals be better served if they were not kept in zoos at all?

CHAPTER 3

Aquariums

Animals that live in or near the water are also featured in exhibits or displays. In fact, you might have such a display in your home or classroom. A container filled with water that houses plants and animals is called an *aquarium* (uh-KWARE-ee-um). (*Aqua* means "water.") In homes, aquariums are usually small glass tanks filled with water and a variety of fish. Those aquariums that house large ocean animals are basically the same as household aquariums, but much bigger.

IN THE BEGINNING

Like zoos, the history of aquariums dates back thousands of years. About 2,000 years ago, ancient Romans were entertained by a variety of swimming creatures that were kept in pools. Aquariums were also popular in ancient China. Here, the typical fish on display was the goldfish.

Although some aquariums were built in Europe during the 1800s, many were destroyed or abandoned during wars. One of the first aquariums in the United States was built in 1873 in Woods Hole, Massachusetts. The New York Aquarium opened its doors in 1896. Most of the aquariums we enjoy today, however, were built in the 1900s.

UNDERWATER ZOOS

People who visit an aquarium usually have an amazing experience. Because people are essentially creatures of the land, it is hard to imagine what the underwater world is like. Many aquariums have found unique ways to present underwater life from the animals' perspective. Visitors may walk by enormously large glass tanks and peer at the animals swimming beyond the glass. A variety of different water animals often live in the same habitat as they would in the wild. Some aquariums even have glass tunnels where people gaze up through a glass ceiling, entranced by the fish that swim overhead.

Aquariums also may include special exhibits, such as a coral reef display or a display that features animals of the deep. Some aquariums focus on specific bodies of water and habitats, such as the Amazon River and the Amazon Rain Forest.

A primary goal of aquariums is to entertain. However, today's aquariums also strive to educate visitors, inviting them to explore the underwater world and learn about the animals that live there.

Modern Mission

The mission of most modern aquariums is reflected in the exhibit policy described by the National Aquarium in Baltimore, Maryland: "The Aquarium strives to blend naturalistic exhibit elements with the most modern interpretive techniques, engaging visitors by focusing on the beauty of the aquatic world and thereby eliciting an emotional response and awakening in visitors the desire to be environmentally responsible." This means that the aquarium wants its exhibits to look as natural as possible. The people at the aquarium hope that the exhibits will make visitors realize how beautiful the underwater world is. They also hope that such beauty will increase the visitors' urge to preserve the world's oceans and wildlife.

LARGER OCEAN ANIMALS

Perhaps the most fascinating residents of an aquarium are the large ocean animals on display. Many aquariums have shark tanks, where visitors enjoy watching the sharks smoothly cruising by. At specified times, people watch a shark feeding, during which several aquarium employees wearing wetsuits and scuba gear venture into the water to hand-feed the sharks.

Dolphin and killer whale shows are also popular attractions. As thrilled audiences watch from seating areas, trainers persuade dolphins and killer whales to perform amazing tricks, such as flips, leaps, tail slaps, fin waves, and even kisses. The spectators marvel that the animals are able to learn and follow the cues of the trainers, and come away in awe of the animals they've watched perform.

WHAT'S THE PROBLEM?

Many people believe that these performances enrich the lives of captive animals; the activities give the animals something to do. However, for many animal activists, these performances are a problem. Considering the welfare of the animals, these people point out that dolphins and killer whales are extremely intelligent, which is why their confinement in aquariums is not in their best interests. If the animals had their way, they would much rather be swimming freely in the ocean than performing on cue for spectators.

Another concern is the way in which the animals arrive at the aquariums. Dolphins and killer whales generally live in family groups called *pods*. Many people believe that the capture of a dolphin or killer whale and its separation from its family must be heartbreaking for the animals—both the one that is captured and the family pod that watches the animal being carried away. The animals may be captured in nets or even

Seal Tricks

Like dolphins, seals and sea lions also have the ability to learn quickly and follow cues. Many aquariums and zoos have shows that feature seals and sea lions doing amazing tricks. Many people feel that these shows encourage humans to appreciate the uniqueness of the animals. Others feel that the shows belittle the real abilities of these ocean mammals.

airlifted out of the water. The peaceful world that they knew is suddenly shattered, to be replaced only by a confining tank, in which they stare at the same scenery each day, and people stare back at them in return.

OF LAND AND SEA

Many aquariums also feature animals that live both on land and in the sea. This trend toward mixing the *terrestrial* (of the land) and the *aquatic* (of the water) reflects the new determination of zoos and aquariums to exhibit ecosystems and/or habitats, not just isolated animals.

For example, polar bears live in the Arctic regions of the world, around the North Pole. Their webbed paws and sharp claws enable them not only to walk across the frozen tundra of the north, but also to swim in the cold polar waters. Many aquariums and zoos try to replicate the polar bears' natural habitat, providing both land and water for the polar bears to enjoy. In fact, it is now illegal for zoos or aquariums to house polar bears in small cages.

Penguins live close to the South Pole, at the opposite end of the world from polar bears. Although they are birds, they cannot fly. Instead, they swim through waters from the Antarctic to the equator, leaping up on land to mate and raise their families. (The *equator* is an imaginary line around the middle of Earth. It is an equal distance from the North Pole and South Pole.) Some aquarium exhibits, like SeaWorld's Penguin Encounter in Orlando, Florida, replicate the penguin's natural habitat of both land and water. Visitors can view the penguins from behind a glass wall as the birds dive into and leap out of the water.

Seals and sea lions also spend time on land and in the ocean. One of the first seal exhibits was at the Lincoln Park Zoo in Chicago, Illinois. It was built in the late 1800s. Like polar bear and penguin exhibits, seal and sea lion exhibits have a mixture of land and water. The animals can often be seen lounging in groups on rocks in the sun, just as they would in the wild.

Although these animals appear to live idyllic lives as they cavort in their new homes, animal activists still question the animals' quality of life. In the wild, their natural territories would be much larger than their aquarium tanks. Many animal activists also criticize the hardships of capture and transportation necessary to bring the animals to the facility. On the flip side, aquarium supporters point out that the animals are safe in their human-made enclosures; they are not in danger of being eaten by natural predators. Another positive is that spectators, upon seeing the animals, will come to have a greater appreciation and understanding of the animals.

Understanding Killer Whales

Although these whales are called "killer whales," they are actually part of the dolphin family. They are the largest of all the dolphins. The "killer" part of the name comes from their desire and ability to eat ocean animals, including seals, sharks, squid, fish, and even whales. For all their ferocity against other ocean animals, however, killer whales have never attacked people.

Killer whales and other marine mammals are kept in tanks much smaller than their natural habitats. Zoo advocates contend that this allows people to better understand and appreciate them.

⊚ LENDING A HELPING HAND

Some aquariums serve another function—they help animals in the wild. Frequently, marine animals get in harm's way. They may become entangled in fishing lines or garbage dumped in the ocean. They may become sick and wash up on beaches, becoming stranded. They may end up in rivers without being able to find their way back to their ocean habitat.

In many cases, an aquarium steps in to lend a hand. Aquariums are staffed with experts on marine life, and an aquarium itself has the facilities necessary to rehabilitate sick or injured animals. For example, the New England Aquarium in Boston, Massachusetts, specializes in rescuing marine animals, such as seals, sea turtles, and even small whales and dolphins, that have become stranded on beaches. Florida's SeaWorld has assisted in the rescue and rehabilitation of a number of ocean animals, including manatees and river otters. Whenever possible, the main goal of any rescue and rehabilitation program is to release a healthy animal back into the wild.

AQUARIUMS AND ZOOS TOGETHER

Although most aquariums do not have a traditional zoo alongside, many zoos are now incorporating aquatic animal exhibits into their parks. New York's Central Park Zoo, for example, has polar bear, penguin, and sea lion exhibits. The Seven Seas exhibit at the Brookfield Zoo in Chicago features dolphins, walruses, seals, and sea lions.

Zoos and aquariums are monitored by the American Zoo and Aquarium Association (AZA). Started in 1924, this group calls for zoos and aquariums to maintain the highest standards in their operations, including hiring qualified staff and properly caring for the animals. Zoos and aquariums must apply for membership, and only when they meet the organization's criteria can they become members. To date, 201 zoos and aquariums have been accredited by the AZA. However, over 2,000 zoos and aquariums do not follow any similar regulations. To see if a zoo or aquarium is accredited by the association, go to its Web site (www.aza.org) and click on "AZA Members."

CHAPTER 4

At the Circus

Circuses can be thrilling spectacles to behold. Human acrobats twirl and fly through the air. Men and women perform amazing stunts while riding horses. Daring animal trainers face seemingly tame elephants, tigers, and lions, requiring these animals to do as they are asked.

What kind of life do circus animals really lead? Is it a better life than they would have if they were left in the wild? Is it a life that they would choose and one in which they thrive? Animal trainers claim that they and their animals have mutual respect for each other. Yet animal activists claim that circus animals are often mistreated in a variety of ways, both physically and emotionally.

THE EARLY DAYS OF THE CIRCUS

Historians agree that the basic idea of the circus—a traveling group of performers, both human and animal—dates back over 4,000 years. Art from ancient Egypt depicts clowns and acrobats performing for an audience. In ancient Greece, also about 4,000 years ago, people performed with bulls, grabbing their horns and flipping over the animals' heads.

Also comparable to circus entertainment were the performances held in outdoor arenas during festivals in ancient Rome, about 2,000 years ago. Spectators watched a variety of entertainment: horses pulling chariots as they raced around rings; animals displayed in cages; fierce battles between animals and people, and animals and animals.

Presidential Support

George Washington, America's first president, enjoyed going to the circus. Some people thought that circuses were silly, perhaps even evil. So circus entertainers began reenacting stories from history and the Bible to entice people to attend. When George Washington began going to circus shows, Americans deemed it was acceptable, as well as educational, to go, too.

Fast-forward 1,000 years to the Middle Ages in Europe. At that time, most people did not see much of the world beyond their own small villages. Traveling circuses provided them with another view of the world. Villagers saw not only amazing human performers, but exotic animals, like dancing bears, as well. People looked forward to these traveling shows, which offered a change of pace from their everyday lives. Over the next few hundred years, circuses entertained people with such acts as jugglers, singers, trained dogs, sword-swallowers, and *contortionists*—people who can bend and twist their bodies into amazing positions.

ANIMAL ACTS

A man named Philip Astley is credited with being "the Father of the Modern Circus." Born in England in 1742, Astley grew up to become an accomplished horseman. In 1770, he opened his own horse-riding school. In the morning, the students would learn how to ride. In the afternoon, the skilled riders would perform for paying crowds. Even Astley joined in the fun, standing on horseback and waving a sword above his head. To add to the performances, Astley hired musicians, acrobats, and rope walkers. The audiences were enthralled. By the time he died in 1814, Astley had eighteen circuses in Great Britain and Europe.

At the same time, another horse act was developing in the United States. In Philadelphia, a man named Thomas Poole performed routines on horseback, and clowns performed between acts. Also arriving in Philadelphia was John Bill Ricketts from Scotland. He brought with him his own circus of performing stars, including horses, rope walkers, tumblers, and clowns.

Television and motion pictures did not exist in the 1700s, of course, and many people had not heard of, much less seen, such animals as lions, elephants, camels, and giraffes. However, merchants who sailed the world's oceans to buy and sell goods had seen them. They began bringing back these exotic animals to Europe and North America.

Circus owners realized that people would pay money to view these amazing creatures from faraway lands, and they added the animals to their circus entertainment. The tradition of including exotic animals in circuses had begun.

RINGLING BROTHERS AND BARNUM & BAILEY

Perhaps the most well-known circus of all, also called "The Greatest Show on Earth," is the Ringling Brothers and Barnum & Bailey Circus. It was originally a collaboration by two men,

On the Move

In Europe, circuses did not usually travel. Because the cities there were firmly established, buildings to showcase circuses were built. The United States, however, was a newer country with less-established cities. To find new audiences, the circuses went on the road, traveling from town to town. Some circuses even traveled by boat on the Mississippi and Ohio Rivers and the Great Lakes in the northern Midwest.

Phineas T. Barnum and James A. Bailey, and five brothers, the Ringling brothers. The Barnum and Bailey Circus and the Ringling Brothers Circus had performed independently of each other, but in 1907, the Ringling brothers bought the Barnum and Bailey Circus. In 1918, they combined the two, creating a circus with 2,500 human and animal performers.

This zebra is one of the many animal performers in the Ringling Brothers and Barnum & Bailey Circus.

People who attend a Ringing Brothers and Barnum & Bailey Circus are given a feast for the eyes. Elephants outfitted in sparkling costumes put their front legs on each other's backs, creating a long chain. Ferocious lions and tigers snarl at trainers, who manipulate them into jumping through flaming hoops. Men and women perform daring feats on horseback as other horses gallop around the ring. Monkeys and chimpanzees, dressed in human clothing, scamper across the circus stage. Although these animal acts are exciting to watch, you might wonder—are the animals really happy to be part of the circus?

ENTER THE ACTIVISTS

The mistreatment of animals in circuses dates back to the circus's earliest days. Animals were routinely injured and killed in the festivals of ancient Rome. However, it would take several hundred years before people began to pay attention to the ways that circus animals were treated.

One of the first advocates for animal rights was Henry Bergh. Living in New York City in the mid-1800s, Bergh became outraged at the mistreatment of horses, cats, and dogs that he witnessed every day. In 1866, he started the American Society for the Prevention of Cruelty to Animals (ASPCA). At first, he and Barnum were constantly at odds, because Bergh championed better treatment and living conditions for circus animals. Eventually, Barnum came to respect Bergh's views on animal welfare, and he even helped start the Connecticut Humane Society.

Today, the plight of animals in circuses is still a concern for animal activists. They are quick to point out that animals such as lions, tigers, and elephants would much rather be in their natural habitats than performing under the big top. In between performances, the animals are caged or chained up, denied any freedom. The cages allotted to these animals are usually just large enough so that the animals can lie down and perhaps pace

The Animal Welfare Act

In 1966, the United States Congress passed the Animal Welfare Act (AWA). Its goal is to "protect certain animals from inhumane treatment or neglect." The law states:

> The AWA requires that minimum standards of care and treatment be provided for certain animals bred for commercial sale, used in research, transported commercially, or exhibited to the public. Individuals who operate facilities in these categories must provide their animals with adequate care and treatment in the areas of housing, handling, sanitation, nutrition, water, veterinary care, and protection from extreme weather and temperatures.

> Basically, the law says that anyone who sells animals, conducts experiments on animals, takes animals that are for sale from place to place, or shows animals to the public (such as in zoos or circuses) must give the animals basic care. This includes providing the animals with a good place to live; keeping them clean; making sure that they have healthy food, water, and veterinary care; and protecting them from bad weather.

a few steps back and forth. This is a stark contrast not only to their natural habitats, but also to zoo conditions, in which animals may have as much as 1 to 2 acres (0.4 to 0.8 hectares) to call home. In addition, animal activists say that the methods used to train the animals to do their tricks are often cruel, including beating the animals into submission.

Ringling Brothers claims that these accusations are not true. "Ringling Brothers does not tolerate the abuse or mistreatment of any animal," the circus Web site states. "Our training methods are based on positive reinforcement in the form of food rewards and words of praise." This means that Ringling Brothers does not allow anyone to harm its animals in order to train them.

Instead, the animals receive food or kind words from their trainers when they do things correctly. Ringling Brothers also points out that it is routinely monitored by the United States Department of Agriculture (USDA) for its compliance with the federal Animal Welfare Act (AWA), which protects certain animals other than those raised for food. PETA, however, lists several instances where Ringling Brothers has been fined by the USDA for its failure to meet the act's standards.

The ASPCA is also unhappy with the circus. The ASPCA is suing the Ringling Brothers circus for its mistreatment of elephants. The society claims that not only are the elephants subjected to beatings and traveling in hot train cars, but sometimes they are chained for up to twenty hours a day. In addition, baby elephants are often removed from their mothers before they are *weaned* (before they have stopped drinking milk from their mothers' bodies). According to the ASPCA, the USDA has explained to the circus that this removal can cause both physical harm and emotional trauma.

Still, the question remains—do humans have the right to train and keep animals for the purpose of circus entertainment? Or do the animals have the right to lead their own lives in their natural habitats?

Roadside Attractions

Ringling Brothers and Barnum & Bailey is one of the largest circuses, but it is not the only circus. In fact, some smaller circuses are also accused of mistreating animals. For example, at least a couple of small circuses have gone bankrupt and abandoned all of their equipment and animals in parking lots. Because the circuses are smaller, they do not draw as much attention from law enforcement or the government as bigger circuses do. Therefore, possible animal abuses by these circuses might not come to the attention of investigators.

CHAPTER 5

Best in Show

Zoos and aquariums usually feature animals that people consider exotic or unusual. In their everyday lives, most people do not see such animals as elephants, giraffes, tigers, polar bears, or seals. People in rural areas, however, might be more familiar with such animals as cows, horses, sheep, and pigs. These animals can often be found on display in the form of competitions. One of the most common forms of animal competition is the livestock show.

AT THE FAIR

State and local fairs are events during which a community comes together to share its successes and celebrate what makes that community special. This usually involves displaying products produced in the community, such as foods, plants, and animals. The animals are usually livestock.

Fairs vary from community to community, as do the animals on display. Many livestock competitions are run by local 4-H organizations. In conjunction with a local cooperative extension service, the 4-H Youth Development Program encourages children to raise and care for animals and then compete against other children who have raised the same types of animals. The children are awarded ribbons and trophies for their efforts—and for their animals. In addition, some children also learn an important life lesson. Sometimes, the animals that children raise are eventually sold at auction to be used for meat.

Children not only learn the importance of good animal care, but they can learn to appreciate and respect animals that are killed for food.

Adult fair-goers may take part in livestock competitions, too. The animals are often displayed in stalls for people to walk by and view. Beef cattle, dairy cows, goats, horses, poultry, sheep, pigs—these may all be part of one fair exhibit and competition.

Although they are entertaining for the visitors who simply view the animals, livestock competitions are also educational for the children who raise and care for the animals they present. Below are some of the goals for children who raise animals, as stated by the North Carolina 4-H Livestock Program. (While similar, chapter programs and policies vary from state to state.)

 To experience the pride of owning livestock and to be responsible for its management

 To learn skills in livestock production and gain an understanding of the business of breeding, raising, and promoting livestock and [its] end products

 To promote a greater love and understanding of animals and a humane attitude toward them

 To increase knowledge of animal by-products and how animal by-product usage touches our lives each and every day

Not all fairs have livestock competitions. In some smaller fairs, animals are displayed as part of petting zoos, in which children and adults can pet goats, sheep, pigs, horses, and donkeys. Other fairs include races with livestock or other common animals, like skunks. Animal activists complain about the stress

What Is 4-H?

4-H is an international organization for children and young adults, ages five to nineteen. It helps children learn new skills and consider different careers. It also encourages children to get involved with their communities. The four H's in 4-H stand for head, heart, hands, and health. These four words are part of the 4-H pledge:

I pledge
My head to clear thinking
My heart to greater loyalty
My hands to larger service
My health to better living
For my club, my community, my country, and my world.

that the animals might endure in petting zoos or in races, as well as in being transported to the fair.

GONE TO THE DOGS

According to a 2001–2002 survey conducted by the American Pet Products Manufacturers Association (APPMA), 40 million U.S. households include a dog as part of the family. Many households have more than one dog, which is why the APPMA estimates that 68 million dogs are owned by people in the United States.

Where do pet owners get all these dogs? The same survey estimates that 20 percent of all dogs—that is, 13.6 million dogs—were adopted into their new families from animal shelters. The remaining number of dogs—54.4 million—came from other sources, such as pet stores, breeders, friends, families, or neighbors.

More than 2,500 dogs compete each year in the Westminster Kennel
Club Dog Show at Madison Square Garden in New York City.
These dogs are shown in the 121st show, held in February 1997.

A dog *breed* is a specific kind of dog. For example, a poodle is a specific breed of dog. Other dog breeds are golden retrievers, German shepherds, and border collies. The American Kennel Club (AKC) recognizes 152 breeds, and the ASPCA estimates that 60 to 65 percent of all dogs owned are purebred. For many people, owning a specific breed of dog is important, for different reasons. Perhaps a small dog is most suitable for someone's living situation. Or perhaps a person had a particular dog as a child and always wanted to own one as an adult. Or perhaps a person just likes the way that a certain breed looks.

For many people, owning a special breed of dog is also a matter of pride. Dog owners who are proud of their dogs like to show them off. One way to do so is at a dog show. Dog shows range from very small local shows to very large national shows. In a dog show, dogs of the same breed compete against each other. A judge walks among the dogs, looking for characteristics that are special to the breed, such as the way the dog stands, the curl of its tail, the tilt of its ears, and so on.

Through training and experience, dogs learn to be patient during the judging process, allowing the judge to run his or her hands over their bodies. Most dogs appear to enjoy the attention, and the experience is usually not as stressful for the dogs as for the owners, who hope that their dogs will win a prize.

The Super Bowl of the dog-show world is the show presented by the Westminster Kennel Club. It is held every year in February at Madison Square Garden in New York City. This is not a dog show for the first-time participant. All the dogs that enter the Westminster Kennel Club Dog Show—2,500 in all—have been leaders in their divisions and won local shows in the past. The Westminster Kennel Club Dog Show is the chance for the top dogs in the country to compete against each other, all vying to be Best in Show.

The dogs first compete against other dogs of the same breed. The winners in each breed then move on to compete in categories. These categories—herding dogs, hounds, sporting dogs, nonsporting dogs, terriers, toy dogs, and working dogs—

Kirby is a male papillon, one of the many dog breeds that compete at the Westminster Kennel Club Dog Show. He won Best in Show at the February 1999 event.

group breeds according to common traits and history. The winner in each category then goes on to the ultimate competition—Best in Show. The judges do not compare the dogs in each category with each other, but rather compare how well each dog measures up to the standard of its breed. In 2001, the winner of Best in Show was a bichon frise named Special Times Just Right. This dog had won first place in the nonsporting category.

The Westminster Kennel Club Dog Show receives a lot of media coverage and is even broadcast on television. For the dogs, it's just another day to strut their stuff. Dogs love to be

The AKC's Objective

The American Kennel Club (AKC) was founded in 1884. Its main objective is to "advance the study, breeding, exhibiting, running and maintenance of purebred dogs." This means that the AKC works to improve the ways that dogs are bred, raised, shown, and given care. To meet that end, the AKC encourages people to obtain their purebred puppies from responsible breeders. The AKC states: "A responsible breeder is the best source for a well-bred, healthy dog. The breeder will carefully select the parents of each litter to emphasize desirable attributes and minimize faults in their progeny. Some people breed dogs only to produce puppies to sell. These individuals have no regard for the advancement of that breed; they are motivated solely by profit."

According to the AKC, a responsible breeder is one that cares more about the dog breed than about the money that he or she can make from selling the dogs.

with their owners, so the travel and preparation necessary for the dog show are not usually a hardship for them. As the dogs' handlers run the dogs around the ring, the dogs appear to prance with excitement themselves. It seems obvious that the dogs enjoy what they are doing.

PUPPY PROBLEM

Most people who wish to show purebred dogs want to know a lot about a dog's history. For that reason, these prospective dog owners will usually purchase their dogs through *dog breeders,* people who breed specific types of dogs. The dogs bred and raised by breeders usually get good care.

*Two yellow Labrador retrievers are shown in holding pens at the
Animal Rescue League in Des Moines, Iowa, in March 2001.
They were among thirty-nine dogs seized from a puppy mill.*

However, many people do not care what a dog's background is. The only thing that is important to them is that the dog is of a specific breed. The easiest place to find a dog of specific breed is usually a pet shop. The problem with many pet shops, however, is that they receive their purebred puppies from what are called *puppy mills*. Puppy mills are facilities that mass-produce dogs for the purpose of selling them, just as one would mass-produce sneakers in a factory.

What concerns animal rights and animal welfare activists is the way that the animals at puppy mills are treated. The puppies are often crammed into cages, receive very little veterinary care, and do not socialize much with humans. The puppies are shipped to pet shops when they are about eight weeks old. Because of the poor conditions in which these puppies live, many of them have health problems.

Although conditions in puppy mills have come under fire since the 1980s, the mills still thrive, mostly because many pet owners insist on owning purebred dogs. This wasn't always the case in the United States. People who own purebred dogs often register their pets with the AKC. In 1944, during World War II (1939–1945), only 77,400 dogs were registered. Five years later, that number had more than tripled to 235,978. Perhaps this was because some people came to equate owning a purebred dog with leading a more financially secure life than they had led during the war. In 2000, 1,175,473 dogs were registered, the most popular breed being the golden retriever.

Other options are available to people who wish to own dogs. The local animal shelter is one place to look for a pet. Here, people might not only find the dog of their dreams, but they can offer an abandoned dog a second chance. Prospective pet owners can also try rescue groups. Organizations dedicated to specific breeds often rescue dogs that have been left at animal shelters. A third option is to adopt a *mutt*—a dog of mixed breeds. Mixed breeds are just as deserving of loving homes as purebreds. And that's the bottom line for animal lovers—that all dogs find loving, permanent homes, whether a dog is registered with the AKC or is of mixed heritage.

Petfinder

For people who really want a particular breed of dog but would like to adopt from a shelter, an online service might be the answer. At Petfinder, you can type in the breed of dog that you would like. The service then lists animal shelters in your area that have that type of dog up for adoption. The Web site address is www.petfinder.org.

CHAPTER 6

At the Races — Greyhounds

Many people enjoy watching things that move fast. Millions of spectators watch race cars as they whip around a track. Speed is the goal for Olympic swimming and track stars. Baseball players run as fast as they can around the bases to score runs. Hockey players zip around an ice rink to score goals. But perhaps nowhere is speed more important than in dog and horse racing, because sometimes the animals are literally racing for their lives.

CHANCES ARE

In dog and horse racing, the animals are required to race against other animals around an oval track. The purpose of these races is not for the spectators to appreciate the beauty and strength of the animals, but for the spectators to *gamble,* or place bets.

When people place bets, they pay money to the racetrack in the hopes of making more money. First, the bettors figure out which dog or horse they think will win a race. Then they pay money to place bets on the potential winners. Experts assign numbers to each animal to tell bettors its chances of winning. For example, if a dog consistently wins races, the chances of that dog winning are very high. If a dog does not win often, the chances of that dog winning are very low.

Gamblers like winners. When a dog wins a race, the people who bet on that dog get more money back than they originally bet. If the dog or horse does not win, the gambler loses the money that he or she has bet. A winning animal can also make

A Matter of Probability

Probability is the mathematical principle behind dog and horse racing. If an animal has a 2 to 1 chance of winning, it means that the animal has a good chance to win. The odds are 1 in 2, or 50 percent, that the animal will be a winner. If the dog or horse has a 17 to 1 chance of winning, it means that the animal is not favored to win. The odds are 1 in 17, or 5.9 percent, that the animal will be a winner.

a lot of money for its owner and trainer, because prizes are given to the winning animals.

However, many people feel that the life of racing animals is not often a good one. Some owners use unethical means to improve the performance of their dogs or horses, such as giving them performance-enhancing drugs. Track officials often test the animals through blood and urine samples to stop this form of cheating. Also of concern to animal activists is what happens after a dog or horse has finished its racing career.

GREYHOUND HISTORY

The greyhound is a specific breed of dog, one of the 152 breeds recognized by the AKC. The greyhound is also one of the oldest breeds of dogs on record. Drawings of greyhoundlike dogs helping humans hunt date back 8,000 years. Famous figures in the history of ancient Egypt—such as King Tutankhamen and Queen Cleopatra—had greyhounds. Some greyhounds were even mummified.

Long ago, greyhounds were used to help people hunt for food. Greyhounds are sight hunters. This means that instead of finding animals by following their scent, greyhounds rely largely on the sense of sight. And greyhounds are fast—very fast. They can run up to 42 miles (67 kilometers) per hour. Their sleek bodies, keen eyesight, and great speed make them ideal hunters.

Long ago, people who owned greyhounds loved to watch them chase after prey. In fact, they loved it so much that they invented a sport for that purpose. The sport was called *coursing*. In coursing, greyhounds would chase after prey, usually a rabbit, on a designated course. The sport of coursing dates back hundreds of years to ancient Greece. Eventually, greyhounds were brought from Greece to Rome, and then into other parts of Europe, such as France and England. Throughout England and France, only royal families were allowed to own greyhounds.

Greyhounds were brought to America by their European owners in the 1800s. This proved useful at the time, for the dogs helped clear the land of small, unwanted animals that dug holes and ate crops, such as rabbits. Greyhounds were also used for coursing, during which they would chase and kill rabbits for fun. As time passed, people began to object to coursing events because of the unnecessary violence toward the rabbits. In 1906, a man named Owen Patrick Smith designed a mechanical device that could move an object for greyhounds to chase. People enjoyed watching the greyhounds race each other as they chased this mechanical object, and the sport of greyhound racing was born.

GREYHOUND RACING

As of December 2000, the National Greyhound Association (NGA) counted forty-eight greyhound racetracks in fifteen states. On a typical track day, a racing program features thirteen races, with eight dogs in each race. People can visit a special pen before each race where the greyhounds are on display.

Racing greyhounds are bred for the specific purpose of being racers. They are born and raised on what are called greyhound farms. Each dog is registered with the NGA.

Greyhounds race around a turn during opening day of live racing at the Dubuque Greyhound Park and Casino in Dubuque, Iowa, in May 1996.

Life for puppies on a greyhound farm is very different from the life of puppies in a home. First, puppies born for the purpose of racing are monitored for their ability to be good race dogs. There have been reports of puppies being *culled*, or selectively removed from a group and killed, because the owners felt that the dogs would not be profitable at the racetrack. Second, greyhound puppies are reared in a kennel environment, and their training is mostly learning how to race and becoming accustomed to the racetrack. Therefore, racing greyhounds are not familiar with many everyday items, such as stairs and windows.

This does not mean that the life of a racing greyhound is poor. Although mistreatment at greyhound farms has been reported, the farms themselves are monitored by the American Greyhound Council (AGC) and the NGA. Because they work with people and other dogs their entire lives, greyhounds are comfortable around people and other animals. What falls into question, then, is what happens to a retired greyhound.

⊚ THE RETIRED GREYHOUND

The numbers tell the story. In 1990, the AGC estimated that 52,000 greyhounds were born. A greyhound's racing career usually ends when the dog is three to four years old. In 1993, 13,000 greyhounds were adopted. What happened to the remaining greyhounds that had been born three years earlier?

In the early 1990s, reports began to surface about the atrocities committed by people against retired greyhounds. When a greyhound's career is over, the dog is often killed. Some greyhounds are painlessly put to sleep. However, others are shot, bludgeoned, and even electrocuted. Estimates of the number of greyhounds killed each year range from 20,000 to 30,000.

Another fate of retired greyhounds is to be used for medical research. Because greyhounds are mild-mannered and obedient, they are perfect specimens on which to conduct experiments. Although the dogs are helping scientists, it is still a sad ending for these gentle animals.

Speed Limits

Here are the speeds of some other animals besides the greyhound.

Animal	Speed
Swift (bird)	100–200 miles (160–320 kilometers) per hour
Cheetah	70 miles (112 kilometers) per hour
Sailfish	68 miles (109 kilometers) per hour
Lion	50 miles (80 kilometers) per hour
Jackrabbit	45 miles (72 kilometers) per hour
Racehorse	43 miles (69 kilometers) per hour
Housecat	30 miles (48 kilometers) per hour
Human	25 miles (40 kilometers) per hour

Erica Lannon of Sharon, Massachusetts, embraces her rescued greyhound Rosie as they join members of the Greyhound Protection League of Massachusetts in March 1997.

GETTING BETTER

National media reports describing the plight of retired greyhounds shocked many people in the United States. Groups sprang up around the country to find lasting, loving homes for these princely dogs. In addition, the number of greyhounds born each year is decreasing, while the number of greyhound adoptions has risen. Many greyhound racetracks have established their own adoption centers. Today, there are over 200 greyhound adoption groups.

The AGC strives to improve the life of racing greyhounds, both before and after their careers on the racetracks. AGC inspectors routinely visit greyhound farms to ensure that the animals are receiving the proper care. The AGC estimates that more than 80 percent of all greyhounds are now either adopted into homes or returned to greyhound farms for breeding. Still, that leaves another 20 percent of greyhounds that are killed unnecessarily.

There is good news, however, for greyhound supporters. According to the American Gaming Association, betting on greyhound races made up only 0.4 percent of gambling revenues in 1997. (Gambling revenues include the money from horse racing, lotteries, casinos, and so on.) Attendance at greyhound races is also down. If this lack of interest in greyhound racing continues, then perhaps someday even fewer greyhounds will be bred for racing. If fewer racing greyhounds are born, then fewer of these greyhounds will need homes after they retire from racing.

Greyhounds by the Numbers

The chart below shows the number of greyhounds born and the number of greyhounds adopted each year. Numbers are from the AGC.

Year	Greyhounds Born	Greyhounds Adopted
1990	52,000	3,500
1991	52,000	7,000
1992	49,000	9,000
1993	43,086	13,000
1994	42,119	14,000
1995	37,650	16,000
1996	36,688	18,000
1997	35,730	18,000
1998	35,801	18,000
1999	33,256	18,000

CHAPTER 7

At the Races—Horses

Greyhounds aren't the only animals that race for people to bet on. Horse racing is also a popular sport in the United States and around the world. In the United States, there are forty-eight greyhound tracks, compared to nearly 100 racetracks for Thoroughbred horses. The basic betting principles in horse racing are the same as in dog racing. People choose which horse they believe will win and then place money to bet on that horse. Racehorses aren't the average horses that you see on a farm. Most horse racing involves Thoroughbreds, which are a specific breed of horse.

THOROUGHBRED HISTORY

Throughout history, humans have challenged each other over the speed of their horses. Horse racing took on new significance in England in the 1600s, when the king, Charles II, began attending races, even providing prizes. With royal approval of racing, people began looking for the supreme horse to win the king's favor.

Something else was happening in England at that time. Traditionally, people rode horses to help them hunt animals in forests. Hunting itself had become a sport for upper-class British society. In the 1600s, however, many forests were being cut down to make room for farms and homes. Hunted animals were now chased across open fields, prompting the need for swifter horses.

Two Thoroughbred racehorses, Affirmed, right, and Alydar, approach the finish line at the Preakness Stakes at Pimlico Race Course in Baltimore, Maryland, in May 1978. Affirmed won the race.

People began looking for the fastest horses possible, and many horses were brought to England from other countries. Andalusian horses came from Spain. Over 100 horses arrived from Arabia and Turkey between 1690 and 1730. In fact, all Thoroughbred horses today can trace their ancestry back to at least one of three individual horses imported at that time: a Byerley Turk, a Godolphin Arabian, and a Darley Arabian. Only horses that can trace their roots back to these three horses are called Thoroughbreds.

The first horse races in England were actually much longer than the horse races of today—ranging from 1 to 4 miles (1.6 to 6.4 kilometers). Two horses raced in a *heat,* or one running of the race. Pairs of horses raced in several heats throughout the day until one horse had won two heats. Sometimes, the horses ran 20 miles (32 kilometers) a day.

With the races came gambling. To hold the attention of the spectators, race coordinators began holding shorter races. This allowed for more races throughout the day. In addition, more horses were added to each race. The seeds of modern racing had been sown.

THOROUGHBREDS LAND IN AMERICA

The first horse races in America were held before Thoroughbreds ever arrived. In 1665, the royal governor of New York opened the first racetrack on Long Island. In the 1700s, Thoroughbreds began arriving in America along with their wealthy British owners.

Horse racing slowly spread across the new country of the United States and gained in popularity. In 1863, the Saratoga Race Course in Saratoga Springs, New York, opened its doors. The track, which is still in operation today, attracted thousands of wealthy spectators, and New York State soon became the center of the racing world. The American Jockey Club, which was formed in 1864, was based in New York City. This group monitors the breeding and racing of horses.

@ A THOROUGHBRED'S LIFE

Most Thoroughbreds are treated well early in life. Thoroughbreds have the potential to make their owners lots of money, so supreme care is taken to ensure that they receive top-quality food, training, and medical attention.

A young racehorse is called a *yearling*. When it reaches a year old, it begins to train. It becomes used to wearing a saddle and responding to a bridle, as well as becoming familiar with the starting stalls of a racecourse.

When a horse is two years old, it is taught how to race. The trainer works with the horse, helping it to develop its optimum racing ability. The horse enters races at this age, and at three years old, it is eligible to race in the most well-known races of all—the Kentucky Derby, the Preakness Stakes, and the Belmont Stakes, the three races that make up the Triple Crown.

Although Thoroughbreds can race to the age of ten, many horses are retired when they are three or four years old. Many male racehorses are then put out to stud, and the females are

People in Horse Racing

- The *owner* is the person who owns the racehorse.

- A *trainer* is a person who trains the horse to race.

- A *groomer* is a person who takes care of the horse's needs, giving the horse food and water, cleaning its stall, bandaging its legs, and washing and grooming its body.

- A *jockey* is a person who rides the horse during a race.

- An *outrider* is a person who rides another horse to lead a racehorse and jockey onto the racetrack. Each racehorse has its own outrider.

used as brood mares. This means that the horses are used to breed other potentially race-winning horses.

⌗ HORSE RACING CONTROVERSIES

The sport of horse racing poses risks and dangers to the animals. Imagine that you weighed 1,000 pounds (450 kilograms). You would expect that your legs and ankles would be big enough to support your weight—that is, they would be much bigger than the legs and ankles that you have now. However, the ankles of a horse are not much bigger than the ankles of a person. Because these small ankles bear the brunt of all that weight, they can be subjected to numerous injuries.

A leg injury to a horse is often fatal, because horses need their legs to survive. They are constantly on their feet, even when they sleep. When a horse injures a leg during a race or training, sometimes the kindest remedy is to kill the horse and end its pain and suffering. Reports estimate that 800 Thoroughbreds die each year in North America due to racing injuries.

Injuries often occur because a horse is really too young to race. In the horse racing industry, the official birth date for all horses born in any given year is January 1 of that year. Therefore, a horse born in June is officially considered one year old on the following January 1, even though it is really only six months old. Its racing training begins when it is considered to be a year old. Although the horse may look mature on the outside, its muscles and bones on the inside may not be strong enough or sturdy enough to bear the rigors of racing.

A second controversy is similar to the greyhound crisis: What happens to retired racehorses that aren't used for breeding purposes? Unfortunately, many of these horses are sold to factories for the purpose of making dog food, as well as meat for human consumption. Although horse meat is not popular in the United States, it is eaten in Europe. However, one

Harness Racing

Another form of horse racing is *harness racing*. During these races, a horse is harnessed to a type of small chariot called a *sulky*. A driver sits in the sulky as the horse (a type originally bred for harness racing, called a *Standardbred*) races around the track. The driver controls the horse with reins and sometimes a small whip.

requirement of horse meat for human consumption is that the horses must arrive at the *slaughterhouse*—the place where animals are killed for their meat—alive. Horses are often shipped to the factories in double-decker trucks. Because the trucks are intended for cattle, they are not tall enough for the horses to stand comfortably. Trips to the factories are not short, some lasting up to thirty-six hours. The last few hours of these horses' lives are often miserable and difficult.

A third controversy surrounds the use of drugs. Racehorses are sometimes given drugs that their trainers feel will make them run more efficiently. One such drug is *furosemide*. This drug was originally given to horses as a treatment against *exercise-induced pulmonary hemorrhage* (EIPH), which is bleeding in the lungs. There are two theories that explain the cause of this condition. Some people feel that the blood gets into the lungs because horses use more oxygen when racing than when they are moving more slowly. This adds pressure to the blood vessels in the lungs, which causes them to burst. Others believe that the blood vessels burst due to the horse's pounding hooves, which send out pressure waves through the horse's body.

Some horse owners noticed that their horses seemed to run better after they'd been given the drug. Soon, other horse owners began giving their horses the drug, too, even if they did not have EIPH. No one is absolutely positive that furosemide does help a horse run faster. Some people argue that it is not even effective against EIPH. The use of furosemide is so controversial that the state of New York prohibits it, as do many countries around the world.

Harness racing is among the traditions of the Neshoba County Fair, near Philadelphia, Mississippi, shown here in August 1997.

Despite the controversies, horse racing is big business, and the horses themselves are a means to make money. According to the American Horse Council, the horse racing industry generates over $100 billion of revenue each year.

OTHER HORSE COMPETITIONS

Thoroughbreds compete in other events as well. In the *steeplechase*, horses race together around an obstacle course. The steeplechase course has barriers for the horses to jump over, like fences, hedges, walls, and even small bodies of water. In *cross-country jumping,* horses race separately around an obstacle course and are given a score based on the time that it takes them to complete the course, as well as their performance. In *dressage* competitions, the horse and rider demonstrate a variety of movements, including turning in a tight circle. Two team sports involving riding on horseback are horseball and polo. In horseball, the players toss a ball similar to a soccer ball back and forth or lift it from the ground, all while on horseback. In polo, players on horseback attempt to hit a ball on the ground with wooden mallets, or *hammers*.

People have been riding horses for thousands of years. It was only natural thousands of years ago for people to test their

Dangers of Horse Competitions

Christopher Reeve is an actor best known for his role in the *Superman* movies. In May 1995, he was injured in a horse racing accident in Virginia, which left him paralyzed. His horse was about to jump a 3-foot (0.9-meter) fence when it suddenly stopped. His hands tangled in the bridle, Reeve flew over the horse's body, and his head hit the ground. This incident illustrates the dangers to riders during horse competitions.

horses against each other to determine which were the best. The thrill of horse racing and competition are embedded in history, and the role of the horse is of utmost importance. However, many animal activists feel that the horse racing industry as it exists today needs to be more heavily monitored so that these beautiful animals are not abused or endangered.

CHAPTER 8

One, Two, Three — Pull!

About 10,000 years ago, people began to plant crops, which led to the first farms. People soon discovered that animals could help with farmwork by pulling carts or plows. Oxen (farm animals related to cows) and horses were commonly used for pulling farm implements. Scientists believe that people first began to tame wild horses during the Bronze Age, about 3,000 years ago. Most of the horses people see today are domesticated.

SOME HORSE HISTORY

Historically, people used horses for a variety of reasons. Riding on horses enabled people to move quickly from place to place. People rode horses when fighting in battles. Horses could pull carts loaded with objects, as well as carriages carrying people. Horses, therefore, were one of the first means of transportation.

Horses used for work were not usually the Thoroughbreds raised for hunting and racing. Workhorses were often called *draft horses*. They were stockier and stronger than racehorses, able to pull heavy weights and work long hours. Some breeds of draft horses are the Clydesdale, the Percheron, and the Belgian.

In the early 1900s, the use of horses on farms and for transportation slowly dwindled as cars and mechanized tractors became more available. It is extremely rare today to see horses used for transportation or farming, except perhaps in remote rural areas. A few horses are still used as carriage horses, a controversial topic for animal rights and animal welfare activists.

A horse-drawn carriage travels through the Independence Hall area of Philadelphia, Pennsylvania, in June 1998. At the time, city officials were proposing to shorten the work week of the horses, in addition to giving them added protection in cold and very hot weather.

CARRIAGE HORSES

The carriage horse trade today exists primarily in cities as a form of transportation designed to entertain tourists. In most cases, a single horse is expected to pull a carriage with up to six people through the city streets. Although the passengers enjoy these carriage rides as a novelty and a unique way to see the city, many people are concerned about the welfare of these horses.

Carriage horses in the city face several potential dangers to their health. In the wild and even on farms and racetracks, horses walk on ground that is suitable for their hooves and feet—ground that is somewhat soft and giving. City streets, however, are hard concrete. Even though the horses don't usually move faster than a trot, cracked hooves and other injuries can occur. To protect a horse's hooves, people attach horseshoes to them. If the horseshoes aren't replaced regularly, however, the result can be a problem with blood circulation known as *navicular disease*. Horses also must pull their carriages in just about any type of weather, hot or cold. In hot weather, horses—like people—run the risk of getting heatstroke.

New York Rules

About 150 carriage horses ply their trade near Central Park in New York City. Laws were passed to ensure the welfare of these horses.

- Horses must not be worked for more than nine hours in a twenty-four-hour period.

- Horses must be given fifteen-minute breaks and fresh water every two hours.

- Horses must not be driven faster than a trot.

- Horses can work only when temperatures are between 18° and 90°F (−8° and 32°C).

- During winter months, horses must be blanketed when waiting for passengers.

- If a horse is hurt, it must immediately stop working. It cannot return to work until it is seen by a veterinarian.

Wild Horses Today

Of the more than 150 horse breeds in the world today, only one horse still remains wild—the Przewalski's (pshuh-VOLL-skeez) horse of central Asia. The horse actually disappeared from its natural habitat in the 1960s and could be seen only in zoos. However, efforts are being made to reintroduce the Przewalski's horse back into the wild.

The traffic, noise, and even the air of the city can also be a problem for the carriage horses. People who support the carriage horse trade claim that the horses are used to the crowded, noisy streets. But even the most mild-mannered horse can be spooked by a sudden noise or a car that zooms too close. The horse's fearful reaction can cause injury not only to the horse, but to pedestrians as well. Many carriage horses wear blinders on the sides of their eyes to lessen their awareness of city distractions. The city environment also subjects the horses to pollution. A horse's nose points downward toward the ground—toward the tailpipes of automobiles. Tailpipes emit fumes from the cars' engines. As a result, horses often breathe these car fumes at fairly close range.

Another problem is the care of the horses in general. According to the ASPCA, the stables in which carriage horses are housed after their workday are often in poor condition, not to mention the fact that the horses lack pastures in which to graze.

In today's world of modern transportation, many animal activists feel that there is no need for the carriage horse industry. Carriage horse owners and operators trying to protect their jobs claim that the activists are overreacting. All people need to decide for themselves if they think the carriage horse business is fair to the animals.

⊚ SLED DOGS

Imagine a land of ice and snow, a land that cars have a hard time traveling over, a land where people can move faster on skis than on foot.

This is the landscape of Alaska, especially in winter. Although gas-powered snowmobiles are a modern way to get around in this frozen climate, some people still prefer the reliability of a sled-dog team. Like horses, the sled dogs are required to pull a load, usually a person standing or sitting on a sled.

It is true that sled dogs appear to enjoy the activity of pulling the sleds through the snow. Many animal activists, however, object to the conditions in which many sled dogs live. Sled dogs are often housed in outdoor kennels, where each dog is chained to its own doghouse. Outside is where the sled dogs remain, constantly chained to their houses day after day, except when training or sledding. Unlike household pets, they are not invited into their owners' homes to sit before the fire or sleep at the foot of the bed. Their sole existence is as sled dogs, living outside with the rest of the sled-dog team.

The sled dogs do receive the basics when it comes to care. They are fed and receive fresh water. They have shelter to avoid rain or snow. They have the companionship of their fellow sled dogs nearby. Controversy surrounds the fact that the dogs are chained, or tethered. The USDA feels that it is inhumane to tether a dog for most of its life. However, the International Sled Dog

The Alaskan Husky

Although many dogs can be sled dogs, the primary breed is the Alaskan husky. It is a mix of other Arctic dogs, like the Siberian husky and the malamute. Alaskan huskies are lighter and swifter than their Arctic counterparts. The Alaskan husky is not a breed recognized by the AKC.

*Five-time Iditarod champion Rick Swenson mushes his
dog team out of Wasilla, Alaska, in the 1996 race. He later had
to withdraw because of the death of one of his dogs during the race.*

Veterinary Medical Association, which lists one of its objectives as "to actively promote and encourage the welfare and safety of the sled dog athlete," supports the tethering of sled dogs.

THE IDITAROD

The Iditarod (i-DIT-uh-rahd) is a grueling race that is run every winter in Alaska across 1,150 miles (1,840 kilometers) of frozen landscape. Teams of sled dogs and *mushers*—the people who travel on the sleds and control the dogs—travel from Nome all the way to Anchorage. The race can take nine days or more, depending on the weather. Racers stop at checkpoints along the way to rest their dogs and themselves.

The Iditarod is particularly controversial to animal activists. They claim that the dogs are forced to run not only over extremely long distances but also in painfully cold and harsh weather. Dogs have died because of the race. The first race was officially held in 1973. During that first race, between fifteen and nineteen dogs died. The *Anchorage Daily News* reported in 1997 that at least 107 dogs had died since the Iditarod began. Between 1998 and 2000, seven more dogs perished.

A Race for Life

Officially, the Iditarod commemorates the historic 1925 "serum run." When several cases of *diphtheria*—a fatal disease—broke out in Nome, Alaska, in 1925, the only town that had enough serum to fight the disease was Anchorage. Starting on the train and ending with a relay of sled-dog teams, the serum reached Nome in only five days, seven-and-a-half hours. Usually, the trip took twenty-five days. The Iditarod race today travels over some parts of the original serum run.

Super Sledder

Susan Butcher is one of the most successful racers in Iditarod history. Within a five-year span, she won the race four times (1986, 1987, 1988, and 1990) and came in second once (1989). She has even taken her sled-dog team to the top of Mount McKinley in Alaska, the highest mountain in North America. Although her last Iditarod was in 1994, Butcher still lives in Alaska and raises sled dogs. People can meet her dogs as part of a tour along the Chena River.

When one sees a sled-dog team out for a leisurely trip across the snow, it is easy to see that the dogs do, indeed, enjoy what they are doing. What comes into question, however, is the ethics involved in forcing the dogs to race over a thousand miles through some of the harshest weather conditions on Earth. Many people feel that the dogs are just doing what they love, while others feel that the dogs are being exploited for entertainment and sport.

TO PULL OR NOT TO PULL?

For most animals, pulling an object is not a terrible hardship. If the animal is fit and strong, it is not cruel to ask the animal to pull. For example, pulling a cart of children across a grassy field during a hayride is a lot less strenuous for a healthy horse than pulling a carriage through crowded, dirty city streets. The question of animal welfare arises when horses are pushed beyond their abilities for the purpose of human entertainment or profit. Some animal welfare activists believe that many animals that pull are worked beyond their physical limits. They feel that these animals deserve better lives than the ones that their human caretakers provide for them.

Pull Competitions

One way in which animals entertain people by pulling is in pull competitions. Here, an animal is strapped to a cart and encouraged to pull it, usually to a finish line. In some competitions, objects are added to the cart to gradually increase the weight. The animal that is able to pull the most weight the longest distance wins the competition. Horses, ponies, oxen, and dogs are commonly used in pull competitions.

Bred for strength and endurance, draft horses, shown here in Indiana in March 1997, remain the basic source of power for Amish agriculture. Many pull competitions trace their origins to this type of farm work.

CHAPTER 9

At the Rodeo

Rodeos arouse strong emotions in people. People who support rodeos see them as a connection to America's past, as a way to hold on to the traditions that helped make the United States. People who don't support rodeos cite the welfare of the animals as their main reason. They feel that the animals are being cruelly treated for the mere purpose of entertaining crowds and winning prizes.

SOME RODEO HISTORY

To understand the purpose behind rodeos, it helps to take a quick look back at the past. Rodeos began in the American West in the 1800s, when cowboys moved cattle across ranches, tamed wild horses, and ran down wayward calves. The cowboys often camped outdoors, far away from towns or other people.

To add some excitement to their lives, they began having contests. For example, who could stay on a bucking wild horse (or *bronco*) the longest? Who could capture and rope a calf the quickest? These casual contests of skill were usually held at the ranches where the cowboys worked.

Other people besides the cowboys were also looking for excitement. There was not much to do in the American West in the 1800s by way of entertainment. Soon, people began to visit the ranches to watch the cowboy competitions. Two towns claim to have held the first rodeo—Prescott, Arizona, in 1864 and Pecos, Texas, in 1883.

Wild West Show

Buffalo Bill, whose real name was William Cody, put on Wild West shows between 1882 and 1920. They were similar to rodeos in that they featured competitions. But Wild West shows also included reenactments of historical events and demonstrations of frontier skills. Although sometimes considered rodeos, they were not rodeos in the truest sense. A rodeo has traditionally been a competition between working cowboys.

Rodeos have since added competitions other than merely bronco riding and calf roping. No matter the changes, though, the rodeo is a sport that features not merely athletes, but working cowboys. It is the only sport in America that was born of an actual job.

RODEOS TODAY

Today, rodeos are organized events with judges who observe not only the cowboys' skills, but the animals' abilities as well. About 1,000 professional rodeos are put on each year in the United States and Canada. In fact, one of the largest rodeos is the Calgary Stampede in Canada. Held each summer, it lasts for ten days. Rodeos attract about 25 million spectators each year.

All rodeos have basically the same competitions. Some competitions are scored by a judge, and others are scored by a clock. The Professional Rodeo Cowboys Association (PRCA) sponsors most of the professional rodeos and addresses animal welfare concerns as well.

BRONC RIDING

Two types of bronc-riding events are part of a rodeo: *saddle bronc riding* and *bareback bronc riding*. In saddle bronc riding, the bronco is led into a *chute,* which is a narrow pen or passage. Here, a saddle is placed on the horse's back. A flank strap is tied around the horse's flanks, directly in front of the horse's hind legs. Horses usually buck when their flanks are touched, so the flank strap coaxes the horse to buck. While the horse is still in the chute, the cowboy slowly sits in the saddle. He grabs tight to the bronc rein with one hand and gives a nod to open the chute gate.

The bronco leaps from the chute and into the rodeo ring, bucking and kicking, trying to throw the cowboy off its back. The cowboy must stay on the bronc's back for eight full seconds, and he cannot touch the horse or his own body with his free hand; only one hand can hold the bronc rein.

After eight seconds, the cowboy jumps from the horse, and the horse usually stops bucking. Judges rate the cowboy's performance, as well as the performance of the horse. If the horse was very active, requiring the cowboy to demonstrate more skill, the score will be higher than if the horse was less active.

Bareback riding follows the same idea. The only difference is that the cowboy does not have the support of a saddle to keep him on the horse's back.

Rodeo Events

Events Judged for Performance	Events Judged for Speed
Saddle bronc riding	Calf roping
Bareback bronc riding	Steer roping
Bull riding	Steer wrestling
	Barrel racing

Amateur cowboy Frank Winters takes his turn at saddle bronc riding during the fifty-eighth annual XIT Rodeo and Reunion in Dalhart, Texas, in August 1994.

WHY DOES A HORSE BUCK?

The controversy over bronc riding arises over the question of why the horses buck. A horse that is wild will buck when something is placed on its back. In the wild, a horse thinks that a creature landing on its back is a *predator,* an animal that wants to eat it. Instinctively, the horse bucks to throw the animal from its back.

In modern rodeos, cowboys do not ride wild horses, but domesticated ones. The rodeos claim that some horses have the urge to buck as part of their instinctive nature. These are the horses used in rodeos. A horse that enjoys bucking produces offspring that will buck, too. So horses are usually bred for the purpose of performing in rodeos for the bronc-riding events.

Animal activists paint a different picture. If the animal is bucking to throw off a predator, then the animal is fearful and fighting for its life. Bronc riders also wear dulled spurs, which they rub across the horse's shoulders. This move irritates the horse, adding to the bucking motion. Animal activists see the use of these spurs as inhumane.

The use of the flank strap also comes into question. Animal activists claim that flank straps are what cause the horses to buck because they are cinched so tightly around the horses' bodies that they pinch the horses' genitals. Rodeo groups, including the PRCA, staunchly deny this accusation. The PRCA claims that flank straps, which are covered with soft fleece, do not cause the horses to buck at all, and veterinarians have defended the flank straps as harmless. Other people claim that horses will buck if their flanks are touched, which is the purpose of the flank straps—to create a tickling sensation, not a painful one.

CALF ROPING

Calf roping is still practiced on ranches as part of a rancher's job. Cowboys gallop after a calf that has wandered from a herd and fling a rope around its neck in order to lead it back. Roping is also necessary for *branding,* which is a form of identification. A hot iron rod with a ranch's logo is placed against the calf's hide, burning the ranch logo onto the calf's body. This way, if the calf does get separated from the herd, other ranchers will know to whom the calf belongs.

In rodeos, a calf is released into a ring, followed a few seconds later by a cowboy on horseback. The cowboy chases down the calf and tosses a rope—or *lasso*—around its neck, yanking the calf off its feet. The other end of this rope remains attached to the horse's saddle. The cowboy then jumps from his horse, ties three of the calf's legs together with another rope, and raises his arms in the air when finished. All the while, the horse remains steady, keeping the first rope tight so that the calf is unable to move while the cowboy is working.

It is hard to watch a calf-roping event and *not* feel sorry for the calf. After all, here is a small, young animal, terrified of the large horse charging after it. Up-close pictures of calf roping sometimes reveal a calf's eyes rolling in fear. Animal activists claim that the event has the potential for serious injuries to the calves, such as twisted necks and broken backs. Rodeo supporters point out, however, that the calves jump up immediately after being roped and trot out of the ring. They also defend roping as part of everyday life on a working ranch.

STEERS AND BULLS

Steers are young male cattle that have been neutered (surgically altered so that they cannot reproduce), and *bulls* are adult male cattle that can reproduce. Steers are involved in two events—

Shain Sproul of Arlee, Montana, roped this calf in 9.8 seconds at the Northern Rodeo Association rodeo at the Montana Fair in August 1999.

Women in the Rodeos

Women do not traditionally perform in the same events as men. In fact, in 1930, rodeo organizations banned women from competing in rodeos because they felt that the rodeo events were too dangerous for women. The only rodeo event in which women can compete is barrel racing, in which women on horseback maneuver their horses around a grouping of barrels as quickly as possible. The Women's Professional Rodeo Association now holds rodeo events with women contestants only, featuring the traditionally male competitions.

steer roping and *steer wrestling*. Steer roping is less controversial, but it still calls into question the welfare of the animal. Here, two cowboys on horseback chase down a running steer. One cowboy tries to lasso the steer's horns while the other tries to lasso the steer's hind legs. If they succeed, the steer is flung off its feet, pulled from the front and the back of its body at the same time.

Unlike the events described so far, steer wrestling and bull riding are not part of everyday ranch responsibilities. Steer wrestling was invented by an African-American cowboy named Bill Pickett around 1900. It requires a cowboy to jump from his horse and wrestle a steer to the ground by grabbing its horns and twisting its head. All four legs of the steer must point in the same direction when downed. The cowboy who accomplishes this feat the fastest is the winner.

Bull riding is comparable to bronc riding. The cowboy sits bareback on a bull and holds on to a rope wrapped around the bull's body. The bull also wears a flank strap to initiate a bucking response. This event is more dangerous than bronc riding, because the bull can charge after the rider once he has left the bull's back. Rodeo clowns often jump into the ring to distract the bull so that the cowboy can retreat safely.

Animal activists argue that neither of these events is necessary to ranch life, nor are they humane ways of treating the animals. In addition, animal activists claim that some rodeos use electricity to shock their animals into bucking. The PRCA admits to this practice but explains that the devices cause a "mild shock, but no injury."

RODEO RIGHTS OR ANIMAL RIGHTS?

Opposing views of rodeos are very different. The cowboys and rodeo organizations feel that the treatment of the animals is equivalent to how the animals are treated on ranches. Ranch life is hard and painful, not only for the animals, but for the cowboys as well. Animal activists claim that the cowboys have a choice, but the animals do not. It is not the choice of a bronco to have a person sit on its back, but it is the choice of the cowboy to subject his body to the horse's fearsome bucking.

Although animal activists cry out against the cruel treatment of the animals, rodeo supporters are quick to point out that only a very small percentage of animals have been injured as the result of rodeo events—one in every 2,000 animals each year.

2000 Professional Rodeo Cowboys Association (PRCA) Injury Survey

Total animal exposures in events: 71,743
Number of rodeo performances: 187
Number of rodeos: 57
Number of injuries: 38
Percentage of injury: 0.00053

CHAPTER 10

And in This Corner...

"And in this corner" are the words an announcer uses when introducing the fighters in a boxing ring. For many people, boxing is the ultimate sport. It pits man against man (or woman against woman), with each person's strength and skill on the line. At the start of each round in a boxing match, the fighters leave their corners, come to the center of the ring, and fight each other, or box. When the bell sounds, the fighters return to their corners to rest and recover, preparing for the next round.

Some people feel that people fighting animals and animals fighting animals are also thrilling sports to watch. The most popular of such battles include bullfights, dogfights, and cockfights. The difference, however, is that people choose to fight against each other or the animals. The animals have no such choice.

HISTORY OF THE BULLFIGHT

Bullfighting is a sport steeped in history. An image painted on a wall in Knossos on the island of Crete shows men and women with a charging bull. One person is grabbing the bull's horns while another is leaping over its back. The painting dates back about 4,000 years. Bullfights were also popular in ancient Rome, about 2,000 years ago. Historians believe that early bullfights did not much resemble the pageantry of the bullfights of today; rather, they were merely cruel events in which the bulls were killed without any skill or form.

An unidentified matador kills a bull in a makeshift bullring in Barrancos, Portugal, in August 1999. The villagers went ahead with the bullfight in spite of a law that forbids killing bulls in bullfights and the protests of animal rights groups.

Bullfights began to more closely resemble modern bullfights around the eighth century. The Moors, a people originally from North Africa who ventured into Spain, made the bullfights more ritualistic than before, with men on horseback confronting and eventually killing the bulls. While the men positioned their horses, other men on foot would wave capes to distract the bulls. Audiences began to find the cape-wavers more interesting to watch than the bullfighters on horseback, and bullfighting as it is conducted today slowly emerged.

FIGHTING THE BULL

The day of a bullfight often begins with the "running of the bulls," or the *Encierro*. This traditional event began because the bulls have to be led from the corrals where they are kept to the fighting arena. Sometime during the 1600s, a group of spectators ran in front of the bulls as they were led through the city. Today, as the bulls rush through the city streets, people still dash in front of them to try to outrun them. Although the bulls are not usually hurt during the run—some people may hit them with rolled-up newspapers—the bulls' horns may gouge people who fall or get in their way.

The image most people have of a bullfight is a daring *matador* wearing a fancy costume waving a cape as a bull charges. This is how a bullfight begins. Although the cape is

Which Countries Have Bullfights?

Bullfighting is often associated with the country of Spain, but other countries hold bullfights, too. In Europe, people can go to bullfights in the south of France and in Portugal. In the Americas, bullfights are held in Mexico, Peru, Ecuador, Colombia, and Venezuela.

usually brightly colored, often red, the bull responds not to the cape's color, but rather to its movement. To make the cape appear as big as possible, it is usually draped along the matador's sword. The challenge to the matador is to get as close as possible to the bull's horns.

The next phase of the bullfight involves the *picadors*. Riding horses, the picadors enter the ring holding *lances*—spears with sharp heads. They lance the bull an average of three times, usually around the neck.

The third phase of the bullfight involves the *banderilleros,* who carry sticks with barbed heads. When stuck into the bull, the barbed heads hold the sticks in place in the bull's hide. The banderilleros try to place the sticks in the bull's shoulders, causing the bull to lower its head.

Finally, it is the matador's turn again, for the final act of the bullfight. By now, the bull senses that it is in real danger, which makes the bullfight more dangerous for the matador. The matador approaches the bull with swirls of the cape in a series of traditional movements. The challenge is once again for the matador to get as close to the bull as possible without showing fear.

This "dance" between the bull and matador lasts for several minutes, arousing excitement in the audience and leading up to the bull's death. Finally, the matador places the sword directly between the bull's shoulder blades, forcing it downward into the bull's body and, ultimately, the bull's heart.

THE BULL'S RIGHTS

It's hard to argue that bullfighting is not cruel. No matter the arena, a sport in which an animal is killed in front of a cheering audience is hard to defend. Yet many people do defend it, mostly in the name of culture and tradition. Bullfighting supporters explain that the spectacle is not a horrible waste of animal life, but a beautiful performance of man against beast. Some may mention that the matador is also in danger. After all, the bulls can weigh more than 1,000 pounds (450 kilograms). Not only are the bulls powerful, but their horns can inflict terrible injuries. It is the bullfighter's courage, skill, and mastery over the animal, along with the bull's own courage and skill, that are being celebrated in the ring.

For many other people, however, bullfighting is just plain cruel. It is forced upon the animals only for the purpose of entertainment. Many groups have spoken out against bullfighting, including the World Society for the Protection of Animals (WSPA) and the Humane Society International (HSI). Yet bullfighting still persists, with people cheering the victories of the matadors.

Woman Matadors

Traditionally, matadors are men. However, some women have been matadors, too. The Spanish word for a female matador is *matadora*. Conchita Cintrón was a matadora during the 1940s, and Maribel Atienzar was one in the 1980s. In the 1990s, Cristina Sanchez won acclaim as a successful matadora.

@ DOGFIGHTS

Considering that 40 million homes in the United States have dogs, it's difficult to believe that some people would prefer their dogs to be fighting machines than loving animal companions. Yet in the world of dogfighting, this is the case.

During a dogfight, two dogs are placed in an enclosed area and let loose to fight each other. The purpose is to see which dog can inflict the most harm. Unlike a boxing match between humans, there are no breaks every three minutes between rounds. There may be unscheduled breaks during the fight if the dogs pause for a moment. At that time, the fight judge may order the dogs to return to their corners to catch their breaths. When ready, the dogs are once again released. If a dog does not have the energy or will to fight again, the other dog is the winner.

Dogfights are not timed events, after which judges award points and a winner is declared. Instead, the dogs fight until only one dog is left standing. The other dog is either too injured or too weak to continue. A fight between two dogs can last anywhere from one to two hours.

The dog most commonly associated with dogfights is the *pit bull,* or, to use its more proper name, the American pit bull terrier. The dog's strong body and powerful jaw make it the ultimate combatant in the dogfighting ring.

Why Dogfighting?

The purpose of a dogfight is, of course, to entertain. Another purpose is gambling. As with horse or dog racing, people place bets on which dog they think will win the fight. Unlike gambling on dog or horse races, however, gambling on dogfights is illegal.

This pit bull was taken to an animal shelter in June 2000 as a result of public outcries against attack dogs following the mauling death of a six-year-old boy in Germany.

Many people feel that pit bulls do not deserve their reputations as overly fierce dogs. These people believe that it is not the dogs that are at fault, but the people who raise the dogs and train them improperly. A visit to an animal shelter often reveals many pit bulls that have been given up by their families.

However, pit bulls, when properly bred and trained, are just as deserving of good homes as other dogs.

In this country, dogfighting is illegal in all fifty states. In fact, it is illegal for someone to even attend a dogfight in most states. However, dogfighting still goes on secretly, making it even more dangerous for the dogs. Because dogfights are illegal, dog owners might be less willing to get proper medical care for their injured dogs. Dogfighting is clearly a form of entertainment that does not consider the welfare of the animals.

COCKFIGHTS

Another form of animal fighting is *cockfighting*. Here, two roosters, or gamecocks, are placed in a pen and encouraged to fight each other. Once again, the fights are staged not only for entertainment, but for gambling purposes.

Cocks square off against each other on the outskirts of Calcutta, India, in January 2001. Cockfighting is legal in India for certain tribal peoples. Knives are often used on the feet of the cocks for a more dramatic and conclusive ending.

In a farmyard, roosters may peck at each other to determine dominance. This is where the term *pecking order* comes from. The birds rarely cause each other serious harm. In cockfights, however, not only are the birds more aggressive, but they are fighting for their lives. To increase the thrill of the fight for spectators, steel blades are often attached to the roosters' legs. These blades are called *gaffs*. In cockfighting, most of the fighting is done with the roosters' legs as the roosters try to rake or stab their opponents. Therefore, the gaffs are designed to cause as much harm as possible. According to the Humane Society of the United States (HSUS), injuries can include "punctured lungs, broken bones, and pierced eyes."

More people support cockfights than dogfights. Some argue that cockfighting has been a traditional form of entertainment in countries around the world, as well as in the United States. Others contend that we eat chickens anyway, and roosters have a habit of pecking each other, so there shouldn't be a problem.

Animal rights and animal welfare groups alike ponder why people feel that such entertainment is even necessary. Why do people find it entertaining to watch animals kill each other? Why do people believe that the animals do not have the same feelings and reactions as humans?

Lawmakers have begun to defend animal rights and animal welfare. Cockfighting is illegal in forty-seven states, and more than half the states have made it illegal to attend cockfights.

CHAPTER 11

Debating the Issue

So what is the proper use of animals in entertainment? Is it okay to use animals for entertainment, even if there is the potential for injury, as in rodeos and races? Is it okay to use animals for entertainment as long as the animals are not harmed at all?

Or is using animals for entertainment ethically wrong, no matter how good the quality of their care might be? If so, why?

ANIMAL RIGHTS

There seem to be two very different ways that people approach the issue of caring for and using animals. Some feel that animals should be allowed the same rights as people. This means that animals should be allowed to live their lives in their natural environments, without exploitation or interference from humans.

For example, WSPA states this general principle: "Animals have biologically determined instincts, interests and natures, and can experience pain. Thus WSPA believes that animals have the right to live their lives free from avoidable suffering at the hands of humans, rather than be used simply as 'raw materials' for the benefit of mankind." What WSPA is saying is that animals have feelings, too—both physically and emotionally. Therefore, WSPA feels that animals should be able to live freely without being harmed or used by people. WSPA is against rodeos, dog and horse races, animal fights, and the use of

In March 2000, protesters denounce the Ringling Brothers and Barnum & Bailey Circus as the animals walk across Thirty-fourth Street to Madison Square Garden for their annual performance in New York City.

carriage horses and sled dogs, because animals used in these activities do experience pain at the hands of humans.

WSPA also says: "WSPA believes that all animals kept by, or under the control of, humans must be kept in circumstances appropriate to their species." This means that people who keep animals should keep them in places that are good for them. Therefore, WSPA is against such forms of entertainment as circuses, some zoos, and some aquariums, especially those in which the animals are confined to cages or in displays that do not replicate their natural habitats.

Finally, "WSPA believes that where the welfare of an animal under human control is in question, then the animal must be given the benefit of the doubt." In other words, if there is any question about whether or not an animal is being treated properly, the answer should always be in favor of the animal. To some people, this statement means that the rights of the animals are more important than the rights of people. However, others might argue that because humans are the ones who impose their will over the animals, the animals' rights need to be considered.

ANIMAL RIGHTS VERSUS ANIMAL WELFARE

Some people feel that the principles put forth by animal rights groups are much too radical. They feel that only through working with animals can people truly understand and learn about them. The National Animal Interest Alliance (NAIA) says, "Animal welfare grows and improves as we learn more and more about animals, their behavior, and their management. Animal rights remains stagnant with its dogma of 'no more animal use ever.'" (*Dogma* is a principle or belief considered to be true by a particular group.) According to the NAIA, only through learning about animals and by using animals can animal welfare become better. The NAIA believes that animal rights activists don't want people to ever use animals, even if using them means that we can learn more about them.

The NAIA believes that it is okay to use animals to fulfill human needs, as long as the animals are treated properly. As stated on its Web site: "Animal welfare requires humane treatment of animals on farms and ranches, in circuses and rodeos, and in homes, kennels, catteries, laboratories, and wherever else animals are kept."

Animal rights advocates claim that this isn't good enough. One such group—PETA—explains its guiding principle: "Founded in 1980, PETA operates under the simple principle that animals are not ours to eat, wear, experiment on, or use for entertainment."

KEIKO'S TALE

Animal rights and welfare gained particular attention when the movie *Free Willy* was released in 1993. The story revolves around the bond between a boy and a killer whale named Willy. At the end, the boy releases Willy from the aquarium tank in which he lives so Willy can return to the ocean.

Many people reacted to the movie's message of animal rights. A campaign was launched to free Keiko, the killer whale that played Willy, from captivity. Keiko was born in either 1977 or 1978, and he was captured in 1979 in northern Atlantic waters, near Iceland. After spending several years in aquariums in Iceland and Canada, Keiko was eventually sold to Reino Aventura, an amusement park in Mexico City, in 1985. After *Free Willy's* release in 1993, a story was published revealing that Keiko was constantly suffering health problems, despite the Mexican aquarium's best efforts. Due to public outcry and intense research and negotiations, Keiko was placed in the care of an environmental group called Earth Island Institute, and in 1996, he was flown to the Oregon Coast Aquarium and then to a protected area in Iceland, where marine biologists are working

SeaWorld and Shamu

One of the most well-known killer whales is Shamu, associated with SeaWorld, which has aquariums in Florida, California, and Texas. On its Web site, SeaWorld explains the benefits of having killer whales as part of its aquarium.

1. Most people do not have the opportunity to observe animals in the wild. In a 1995 Roper Poll, 87% of those interviewed agreed that visiting zoological facilities was their only opportunity to see wild animals such as killer whales. The unique opportunity to observe and learn directly from live animals increases public awareness and appreciation of wildlife.

2. In the same Roper Poll, 92% of those questioned agreed that zoological parks are a vital educational resource. In the past several decades, marine life parks have learned a great deal about killer whales from ongoing research programs. The advantages of studying killer whales in controlled areas include the possibility of continuous observations without being impeded by weather, darkness, or location.

with him to prepare him for life in the wild. Keiko even has supervised trial releases in order to socialize with pods of killer whales in the wild. But as of September 2001, Keiko was still not ready to live without human intervention. He seems more comfortable in his human-made surroundings, which is understandable, since Keiko is used to associating with humans.

Free Willy also brought aquariums in general under the spotlight, especially those that house killer whales. In the wild, killer whales live in family groups; yet in aquariums, they often live alone. The question was raised: Are aquariums in the best interest of killer whales?

Keiko the killer whale, star of the Free Willy *movies, swims in his tank at the Oregon Coast Aquarium in Newport, Oregon, in January 1998.*

It's a question that can be asked regarding every zoological park, as well as all other means of using animals for entertainment. Keiko the killer whale obviously entertained audiences in the movie *Free Willy,* provoking greater attention to the plight of animals confined by humans. Is this type of entertainment okay, because the animal educated people? Or was Keiko merely being exploited?

A PERSONAL DECISION

Some ways in which animals are used for entertainment are more difficult to accept than others. For example, watching a dog perform on a television sitcom or pose in a dog show seems acceptable to many people. However, watching a calf being yanked off its feet by a rope around its neck during a rodeo may make some people feel uncomfortable. Some may have no qualms when watching elephants in a natural setting at a zoo, yet shake their heads when elephants are led into a circus ring.

Coming to terms with animal rights versus animal welfare is a personal decision that can be arrived at only by researching the ways in which animals are used for entertainment. For example, reading about the pros and cons of carriage horses might have little meaning, but actually seeing the carriage horses might sway a person's opinion in either direction.

The HSUS, the largest "animal-protection organization" in the country, states the following on its Web site: "The HSUS was founded in 1954 to promote the humane treatment of animals and to foster respect, understanding, and compassion for all creatures." With such respect, understanding, and compassion comes the ability to imagine how the animals feel. Are the animals happy in their zoo environment? Are the animals happy performing on television, doing tricks in a circus, or pulling a sled? Or would the animals rather be in their natural habitats?

Animals do have feelings. They experience emotions and pain. Recognizing these experiences from the animals' point of view is a step in realizing which types of animal entertainment benefit both people and animals.

Glossary

activist a person who tries to change government or social policies

advocate a person who supports or defends a cause

animal rights moral or legal claims of animals

animal welfare the general well-being of animals

aquatic of the water

breed a special type of animal or plant

bronc or bronco a horse that bucks

cohabit to live together

conservation the controlled use and protection of natural resources, like forests and animals

coursing a race in which greyhounds chase after prey; in modern coursing events, the "prey" is usually an artificial lure, like a fake rabbit

dog breeder someone who breeds and raises a specific breed or type of dog

dogma a principle or belief considered to be absolutely true by a particular group

domesticated animal an animal that, through generations, has become used to living with people; its offspring will most likely be domesticated

empathy the ability to understand the thoughts and feelings of others

entertainment something that holds the attention of or amuses others

furosemide a controversial drug given to racehorses to treat exercise-induced pulmonary hemorrhaging (EIPH), or bleeding in the lungs

Iditarod a race across Alaska involving sled dogs

inhumane cruel; lacking compassion

kennel a place where dogs are bred, boarded, or trained

livestock farm animals raised for human use or profit

menagerie a collection or exhibition of live animals

mixed breed an animal whose ancestors are of many different breeds or kinds

natural habitat the place or area in which an animal lives in the wild

naturalistic exhibit a zoo display in which animals live in habitats that closely resemble their habitats in the wild

preservation protection from harm

primates a group of animals that includes apes, monkeys, and humans

primatologist a scientist who specializes in the study of primates

puppy mill a facility that mass-produces dogs for the purpose of selling them

purebred an animal that has many generations of ancestors of the same breed or kind

slaughterhouse a place where animals are killed for their meat

tame animal an animal that lives naturally in the wild but has learned to live with people; its offspring will most likely be wild

terrestrial of the land

Thoroughbred a specific breed of horse used in racing

wild animal an animal that lives naturally in the wild

zoological garden a place where people view and observe animals

zoology the study of the animal kingdom

Bibliography

 BOOKS

Alter, Judith. *Rodeos: The Greatest Show on Dirt.* New York: Franklin Watts, 1996.

Bekoff, M. *Minding Animals: Awareness, Emotions, and Heart.* New York: Oxford University Press, 2002.

Curtis, Patricia. *Animals and the New Zoos.* New York: Dutton, 1991.

Curtis, Patricia. *Aquatic Animals in the Wild and in Captivity.* New York: Lodestar Books, 1992.

Edelson, Edward. *Great Animals of the Movies.* Garden City, NY: Doubleday, 1980.

Granfield, Linda. *Circus: An Album.* New York: DK Publishing, 1998.

Johnston, Ginny, and Judy Cutchins. *Windows on Wildlife.* New York: Morrow Junior Books, 1990.

Knotts, Bob. *Equestrian Events.* New York: Children's Press, 2000.

Koebner, Linda. *Zoo Book: The Evolution of Wildlife Conservation Centers.* New York: Tom Doherty Associates/A Forge Book, 1994.

Sherman, Josepha. *Bronc Riding.* Chicago: Heinemann Library, 2000.

 WEB SITES

American Greyhound Council (AGC)
www.agcouncil.com

American Humane Association (AHA)
www.americanhumane.org

American Kennel Club (AKC)
www.akc.org

The American Society for the Prevention of Cruelty to Animals (ASPCA)
www.aspca.org

American Zoo and Aquarium Association (AZA)
www.aza.org

Brookfield Zoo
www.brookfieldzoo.org

Calgary Stampede
www.calgarystampede.com/stampede

The Humane Society of the United States (HSUS)
www.hsus.org

Iditarod.com
www.iditarod.com/index.shtml

International Sled Dog Racing Association
www.isdra.org

International Weight Pull Association
www.eskimo.com/~samoyed/iwpa

The Jockey Club
home.jockeyclub.com

National Animal Interest Alliance (NAIA)
www.naiaonline.org

The National Aquarium in Baltimore
www.aqua.org

National Greyhound Association (NGA)
nga.jc.net

National Museum of Racing and Hall of Fame
www.racingmuseum.org

Oregon Coast Aquarium—Keiko News Central
www.aquarium.org/keiko/index.htm

People for the Ethical Treatment of Animals (PETA)
www.peta-online.org

Petfinder
www.petfinder.org

Philadelphia Zoo: America's First Zoo
www.phillyzoo.org

Professional Rodeo Cowboys Association (PRCA)
www.prorodeo.com

Ringling Brothers and Barnum & Bailey Circus
www.ringling.com/home.asp

San Diego Zoo/San Diego Wild Animal Park
www.sandiegozoo.com

SeaWorld/Busch Gardens Animal Information Database
www.seaworld.org

WCS—Congo Gorilla Forest at the Bronx Zoo
www.congogorillaforest.com

The Westminster Kennel Club
westminsterkennelclub.org

The Wildlife Conservation Society (WCS)
www.wcs.org

Women's Professional Rodeo Association
www.wpra.com

World Society for the Protection of Animals (WSPA)
www.wspa-international.org

Zoo Atlanta
www.zooatlanta.org/main.html

SOURCES

www.AHAfilm.org/history.html

wcs.org/template.php?node=12

www.aqua.org/information/whoweare/mission.html

www.aphis.usda.gov/oa/pubs/awact.html

www.ringlingbrothers.com/menagerie/animal_care

www2.ncsu.edu/unity/project/www/ncsu/cals/
an_sci/ann_rep94/mccla13.html

www.4-h.org/info/whatis.php3

www.akc.org/insideAKC/mission.cfm

www.akc.org/love/dah/howtofind.cfm

www.wspa-international.org/aboutus/policy4.html

www.naiaonline.org/body/animal_welfare.htm

www.seaworld.org/infobooks/KillerWhale/conservationkw.htm

www.hsus.org/about/mission.html

Index

Note: Page numbers in italics indicate illustrations and captions.

119